T0341216

Insight on Genotoxicity

Insight on Genotoxicity

Shiv Shankar Shukla, Ravindra Kumar Pandey,
Bina Gidwani, and Gunjan Kalyani

CRC Press
Taylor & Francis Group
Boca Raton London New York

CRC Press is an imprint of the
Taylor & Francis Group, an **informa** business

First edition published 2021
by CRC Press
6000 Broken Sound Parkway NW, Suite 300, Boca Raton, FL 33487-2742

and by CRC Press
2 Park Square, Milton Park, Abingdon, Oxon, OX14 4RN

© 2021 Taylor & Francis Group, LLC

CRC Press is an imprint of Taylor & Francis Group, LLC

Library of Congress Cataloging-in-Publication Data
Library of Congress Control Number: 2020942808

ISBN: 9780367473372 (hbk)
ISBN: 9781003034964 (ebk)

Typeset in Times
by codeMantra

"To my father, who told me the stories that matter."

Late Shri R. S. Shukla

I love to wear my daddy's shoes though my feet are small,

When they are in my daddy's shoes I feel 10 feet tall,

Someday, I will grow up to fill them,

I only hope to be,

As fine a man and great as Dad as my Daddy is to me.

Dr. Shiv Shankar Shukla

Contents

Preface

I am delighted to write the preface for the first edition of "*Insight on Genotoxicity*". We recognised the need for this book because genotoxicity and mutagenicity analyses play a noteworthy role in the detection of perilous effects of therapeutic drugs, cosmetics, agrochemicals, industrial substances, food additives, natural toxins and nanomaterials for regulatory purposes. Evaluation of mutagenicity or genotoxicity with *in-vitro* and *in-vivo* methodologies exerts assorted genotoxicological endpoints, such as changes in structure and number chromosomes and point mutations.

This book examines the impact of ICH guidelines on genotoxicity testing, which is a regulatory requirement for drug discovery and development. Genotoxicity mechanism has also been discussed in detail with a wider perspective.

Genotoxicity testing has been discussed with subsections taking into consideration the performance enhancement of *in-vitro* genotoxicity testing. This book defines appropriate strategies about advances in *in-vivo* genotoxicity testing, which have been listed along with progress and future prospects. High-throughput approaches for genotoxicity testing in drug development and the associated recent advances are discussed.

The computational prediction of genotoxicity with consideration of mutagenicity, chromosomal damage caused and strategies for computational prediction in drug development have been discussed in detail.

The *Allium cepa* test which is a bioindicator of genotoxicity has also been discussed in detail. In the last section, genotoxicity assessment of nano-sized material/particles has been mentioned as a part of recent advancement.

This book will be a backbone in understanding the genotoxicity concept.

Raipur
Dated: April 2020
Dr. Shiv Shankar Shukla

Acknowledgement

First of all, I take this opportunity to offer devotional prostration at the feet of the almighty and my beloved parents.

During the course of writing this book, I got help, advice, suggestions and cooperation from many people. Some advice influenced, inspired and helped me in completing this book. A task becomes easier when you get moral support and encouragement, and if these names are not listed, the project work and the thesis would remain incomplete.

First of all, I sincerely thank the publisher, CRC Press, Taylor and Francis Group, for their kind gesture and cooperation in publishing this work and their painstaking efforts to bring out this book in the present format.

I am highly indebted to the co-authors Dr. Ravindra Kumar Pandey, Dr. Bina Gidwani and Mr. Gunjan Kalyani for their generous help, teamwork, collaboration, cooperation and criticism. I express my thanks for their encouragement to take up this academic venture for the cause of education.

I express my gratitude towards Ms. Renu Upadhyay, Commissioning Editor (Chemical and Life Sciences) CRC Press, Taylor & Francis Group for her meticulous, conscientious efforts in channelling throughout right form the submission of the proposal, to the preparation of the book in its present format till its final submission.

I profoundly thank Ms. Jyotsna Jangra, Editorial Assistant, CRC Press, Taylor and Francis Books India Pvt. Ltd for her assistance in the preparation of this book.

I extend a note of thanks to the Columbia Institute of Pharmacy for providing a platform.

I earnestly thank the management of Jan Pragati Education Society (JPES) Raipur for their active help and cooperation.

Raipur
Dated: April 2020
Dr. Shiv Shankar Shukla

Authors

Dr. Shiv Shankar Shukla gained his Ph.D from University Institute of Pharmacy, Pt. Ravi Shankar Shukla University. Presently, he is Professor at Columbia Institute of Pharmacy. He has more than 70 publications to his credit in reputed journals. He has authored two books titled *"Inflammation: Natural Resources and Its Applications"* and *"Finger Printing Analysis and Quality Control Methods of Herbal Medicines"*. He has written three book chapters published in reputed Publishing House. He is recipient of 7th Young Scientist of Chhattisgarh award from CCOST in 2009 and Dr. P. D. Sethi Annual Award" for the best paper in 2010. He is recipient of Best Oral Paper award in International Conference at International Conference at AIMST University, Malaysia.

Dr. Ravindra Kumar Pandey gained his PhD from the University Institute of Pharmacy, Pt. Ravi Shankar Shukla University. Presently, he is a professor at the Columbia Institute of Pharmacy. He has more than 65 publications to his credit in reputed journals. He has authored two books: *"Inflammation: Natural Resources and Its Applications"* and *"Finger Printing Analysis and Quality Control Methods of Herbal Medicines"*. He has written three book chapters in published in books from reputed publishing houses.

Dr. Bina Gidwani gained her PhD at the University Institute of Pharmacy, Pt. Ravishankar Shukla University. Presently, she is serving as an associate professor at the Columbia Institute of Pharmacy. She has more than 55 publications to her credit. She has authored one book and two book chapters in books from reputed publishing houses.

Mr. Gunjan Kalyani is Post Graduate from Chhattisgarh Swami Vivekanada Technical University (CSVTU), Bhilai, Chhattisgarh. Presently, he is Assistant Professor at Columbia Institute of Pharmacy. He has 29 publications to his credit in various journals of International and National repute.

1 Introduction to Genotoxicity

INTRODUCTION TO GENOTOXICITY

In genetics, genotoxicity is used to depict the substances that have a destructive effect on the genetic material of the cell (DNA, RNA), which affects the integrity of the cell. Genotoxins are mutagens that cause genotoxicity leading to the damage to DNA and chromosomal material, which ultimately results in a mutation. Genotoxins include chemical substances as well as radiation. Thus, the branch of science that deals with the study of agents or substances that can damage the cell's DNA and chromosomes is referred to as genetic toxicology. Oftentimes, genotoxicity is mistakenly confused with mutagenicity. Thus, it is important to note that, *"All mutagens are said to be genotoxic but all genotoxic substances are not mutagenic"* [1].

Genotoxins can be classified depending on their effects [1]:

a. Carcinogens (cancer-causing agents)
b. Mutagens (mutation-causing agents)
c. Teratogens (birth defect-causing agents)

The damage to the genetic material of somatic cells leads to malignancy (cancer) in eukaryotic organisms. Whereas the genetic damage of the germ cells leads to heritable mutations which cause birth defects (Figure 1.1).

Mutations are inclusive of duplication, insertion or deletion of genetic information. These can cause a diverse range of problems in the host, ranging from a wide variety of diseases to cancer [1]. One of the finest ways to control the dent due to mutagens and carcinogens is to categorise the substance or chemical, i.e. anti-mutagens/anti-clastogens (these suppress or inhibit the process of mutagenesis directly by acting on the cell mechanism) and de-mutagens (these destroy or inactivate the mutagens either partially or fully which affects lesser cell population) from the medicinal plants so that they can be utlised as anti-mutagenic and anti-carcinogenic food or drug additives [2].

IMPORTANCE OF GENOTOXICITY STUDIES

Genotoxicity studies enable the identification of hazard concerning DNA damage and fixation [8]. Genetic change plays a partial role in the complex process of heritable effects and malignancy, including the fixation of the DNA damage caused by gene mutation, large-scale damage to chromosomes, as well as recombination and numerical chromosomal changes. These tests play a crucial role in predicting whether the compound has the potential to cause genotoxicity and carcinogenicity [2].

1

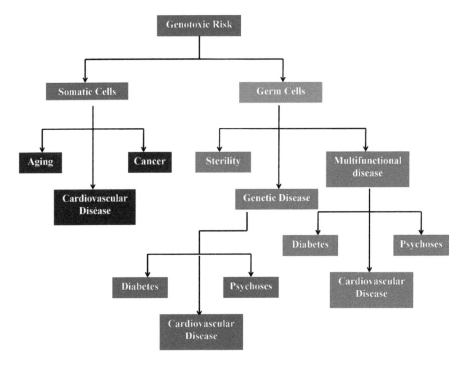

FIGURE 1.1   Risk of genotoxicity.

Regulatory authorities around the globe publish information on the genotoxic potential of newer drugs as a part of their safety evaluation processes. It is usually evaluated along with other toxicological endpoints determined during the safety assessment [2] (Figure 1.2).

During the early testing stages, the same testing assays are performed for predicting both the potential heritable germ cell damage as well as the carcinogenicity because these endpoints have common precursors. The link between exposure to particular chemicals and carcinogenesis has been described, whereas such association has been difficult to ascertain for heritable diseases. Genotoxicity studies have thus been mainly used for the prediction of the carcinogenicity of a compound [3].

CLASSIFICATION OF CARCINOGENS

European Union (EU) classification [4]
a. Carcinogen category 1; causes cancer in humans.
b. Carcinogen category 2; causes cancer in animals and most probably in humans as well.
c. Carcinogen category 3; possibly carcinogenic.
     (Note: evidence for supporting carcinogenicity is inadequate for its classification to category 2).

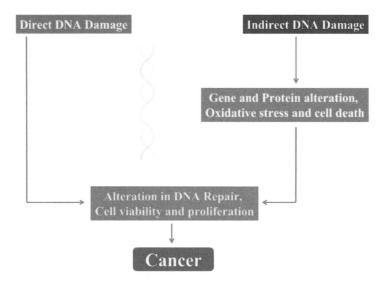

FIGURE 1.2   Relation between genotoxicity and carcinogenicity.

International Agency for Research on Cancer (IARC) classification [4]
a. IARC class 1; carcinogenic to humans.
b. IARC class 2A; probably carcinogenic to humans.
c. IARC class 2B; possibly carcinogenic to humans.
d. IARC class 3; not classifiable as to its carcinogenicity to humans.
e. IARC class 4; probably not carcinogenic to humans.

AGENTS CAUSING DIRECT OR INDIRECT DAMAGE TO THE DNA

Reactive oxygen species (ROS) are genotoxic in nature. Any chemical or substance that increases ROS production adds to the endogenously produced ROS, which, in turn, leads to the non-linear relationships of the dose–effect curve [4]

a. Electrophilic species that form covalent adducts to the DNA.
b. ROS.
c. Ultraviolet and ionising radiations.
d. Nucleoside analogues.
e. Topoisomerase inhibitors.
f. Protein synthesis inhibitors.
g. Some herbal plants such as Aconite, Alfa alfa, Calamus, Aloe vera, Isabghol, etc.

Anti-mutagens decrease the effect of spontaneous and induced mutations. The mechanisms involved are as follows:

a. Desmutagenesis in which factors on the mutagens are somehow inactivated.

   b. *Bioanti-mutagenesis* – these factors act on the process of mutagenesis or repair the damaged DNA, which results in the decreased frequency of DNA mutation [15]. Our cells have DNA repair systems which control DNA mutations naturally.

Major pathways of cell repair for damaged DNA are: [5–7]

   a. Direct repair
   b. Base excision repair (BER)
   c. Nucleotide excision repair (NER)
   d. Mismatch repair
   e. Single-/double-strand break repair

MOLECULAR MECHANISMS INVOLVED IN PRODUCTION OF CHROMOSOMAL ABERRATIONS

One of the endpoints of genotoxicity is gene mutations. Mutagenic chemicals are predominantly responsible for the gene mutations, which are not lethal but can play out as a major threat to the authenticity of chromosomes and viability of cells. The cells are equipped with several DNA repair systems. Depending on the specific classes of DNA lesions, more than one DNA repair pathway becomes active.

   Four of the five key DNA repair pathways that are involved in the repair of DNA lesions foremost to gene mutations are as discussed in detail.

DIRECT REPAIR

Direct repair acts by eradicating or retreating the DNA lesions by a single enzyme reaction in an error-free manner and with elevated substrate specificity. Direct repair does not entail a template as the damage they repair occurs only in one base and there is no engrossment of incision of the sugar-phosphate backbone or base excision. These lesions can transpire due to alkylating agents. Direct repair involves explicit enzymes called alkylguanine DNA methyltransferases (AGMT), which eliminate the alkyl group from the guanine residue of DNA and relocates it to one of its cysteine residues. Next to AGMT, in bacteria and yeast, photolyases can directly reverse ultraviolet-induced DNA damage [6].

BASE EXCISION REPAIR (BER)

BER is a cellular mechanism that repairs damaged DNA through the cell cycle. This mechanism guards cells against the detrimental effects of endogenous DNA damage induced by hydrolysis, ROS and other intracellular metabolites. It is also responsible for the amputation of many lesions induced by ionising radiations and sturdy alkylating agents. The foremost enzymes involved in BER are DNA glycosylases and AP endonucleases.

   The DNA glycosylases are involved in the expurgation of the damaged base, where after the remaining basic site is additionally processed by AP endonucleases.

BER proceeds with the following steps

- short-patch repair (where a single nucleotide is replaced)
- long-patch repair where 2–10 nucleotides are replaced [7]

NUCLEOTIDE EXCISION REPAIR (NER)

NER is a repair pathway implicated in the removal of several kinds of DNA lesions, which mainly begin from exogenous sources such as ultraviolet light or genotoxic chemicals producing colossal adducts and DNA cross-links [7].

NER consists of two different subpathways:

- Global genome repair (GGR) and
- Transcription-coupled repair (TCR).

These two subpathways are unlike the first step of DNA damage detection. The first pathway (GGR) purges overall DNA damage present in the genome. DNA recognition is done by a complex of protein factors (XPC-HR23B and XPE). The second pathway (TCR) removes lesions from active genes.

Here, the crucial trigger in the DNA damage recognition is a stalled RNA polymerase II, which is conveyed by Cockayne syndrome (CS) proteins [8]. The subsequent stages involved in DNA repair are mostly studied for GGR but are indistinguishable for the TCR pathway. After fastening of the XPC-HR23B complex to the damaged DNA in GGR, numerous other proteins are bound, such as a complex called transcription factor IIH (TFIIH) and the endonuclease XPG.

TFIIH includes two DNA helicase activities with reverse polarity (XPB and XPD) that loosen the DNA duplex. After binding of the replication protein A (RPA), the dent is confirmed by XPA where after the endonucleases XPG and ERCC1/XPF slices the 3 and the 5 of the DNA lesion. These grades in the release of a piece containing the DNA damage of 27–30 nucleotides. The enduring gap is crammed in by a complex formed by DNA polymerase d or e, the additional replication proteins, such as the proliferating cell nuclear antigen (PCNA), RPA and the replication factor C [9].

MISMATCH REPAIR (MMR)

Mismatch repair (MMR) identifies and repairs incorrect insertions and misincorporation of bases, which can occur during DNA replication. MMR is a strand-specific repair. During DNA synthesis, the lately synthesised (daughter) strand may include incorrect bases, for example, G/T or A/C.

To repair these incompatible base pairs, it is very imperative to discriminate between the newly synthesised (mismatched) strand and the parental strand.

- The foremost step in MMR is the recognition of the deformity caused by the mismatch.
- The template and the non-template strand are dogged, and the erroneous incorporated base is excised and replaced with the correct nucleotide.

During the repair process, the mismatched nucleotide is removed. A few or up to thousands of bases of the newly synthesised DNA strand can be removed and replaced [10].

CHROMOSOMAL ABERRATIONS AND REPAIR

The one-time endpoint of genotoxicity, chromosomal aberrations, is caused by clastogenic chemicals. Chromosome aberrations can either be structural (which is also referred to as clastogenic) or numerical (which is also referred to as eugenic). DNA damages such as double-strand breaks (DSBs) impact the veracity of chromosomes and the practicability of cells. Unrepaired or misrepaired DSBs can direct to mutations, chromosome rearrangements, cell death and cancer [11–14].

Numerical chromosomal aberrations (aneuploidy) can either trounce or increase the number of chromosomes per cell (like trisomy 21 in Down's syndrome), and it can be deadly or cause genetic diseases. Fortunately, systems to repair DSBs exist. In mammalian cells, DSBs are largely repaired by either homologous recombination repair mechanism (HRR) or non-homologous end joining (NHEJ) repair mechanism [15].

The main difference in HRR and NHEJ is the obligation of a homologous DNA sequence in HRR, which is, therefore, an error-free mechanism. In contrast, NHEJ, which does not utilise sequence homology, is an error-prone mechanism [16]. Another major distinction is their dependence on the cell cycle. HRR depending on the presence of an intact sister chromatid is further competent during late S and G2 phase of the cell cycle when sister chromatids are energetic in dividing cells.

NHEJ does not depend on a homologous DNA strand and can repair DSBs in all stages of the cell cycle, that is, G1, S and G2 phase [16]. HRR acts at the embryonic stage where the embryonic cells are perceptive towards ionising radiation; however, its action in adults is not pronounced unless NHEJ is disabled. Thus, it can be concluded that the input of HRR and NHEJ can be at variance depending on the mammalian developmental stage (which is cell type) and on the explicit type of DNA damage.

HOMOLOGOUS RECOMBINATION REPAIR

HRR is an error-free repair system. The RAD52-group of proteins, RAD50, RAD51, RAD52 and RAD54, and MRE11, play a key role in HRR. In the case of a DSB, the initial cellular response is the breakthrough of the RAD50/MRE11/NBS1 complex. Consequently, followed by a nucleolytic dispensation of the kaput ends of DNA into 3-end single-stranded DNA.

The single-stranded DNA is bound by RPA (replication protein A). After RPA is separated and substituted by RAD51, the RAD51 nucleoprotein filament intervenes in the search for a homologous duplex template of DNA where after the merging of joint molecules amid the broken DNA ends, the integral ds DNA repair template is formed.

The Rad52, Rad54, Rad50 paralogues (such as Rad51B, Rad51C, Rad51D), Xrcc2, Xrcc3 and Dmc1 also play a role at various stages of HRR. After polymerisation of nucleotides to repair damage DNA strands and decree of the recombination intermediates, the HRR is completed resulting in a fault-free double-stranded DNA.

The proteins BRCA1 and BRCA2 involved in breast cancer susceptibility are concerned in HRR as well; however, their role is not well very well understood. Loss of all HRR factors can result too early or mid-embryonic lethality in mice [16].

Therefore, HRR plays an imperative role in improvement, presumably to repair spontaneously arising DNA damage, which is consistent with the findings that HRR and NHEJ can play diverse roles during the mammalian developmental stages [16].

NON-HOMOLOGOUS END JOINING REPAIR

NHEJ is an error-prone repair mechanism. There are at least three steps involved in NHEJ.

- The first step is the revealing of the strand break and the end binding, which is mediated by DNA-PK consisting of three subunits DNA PKCS (DNAb-dependent protein kinase catalytic subunit) and the KU80/KU70 heterodimer, which are implicated in the formation of a molecular bridge that keeps the broken DNA together.
- The NBS1/MRE1/RAD50 complex involved in the processing procedure modifies the non-matching and/or damaged DNA ends into irreconcilable and ligatable ends.
- Finally, in the ligation step, a complex consisting of DNA ligase IV and XRCC4 (X-ray repair cross-complementing) flawed revamp in Chinese hamster mutant ligates the two DNA ends with each other forming an intact double-strand DNA molecule [17].

Recently, Cernunnos XLF was reported, which is also implicated in NHEJ [17]. Cernunnos XLF interrelates and excites the DNA ligase IVXRCC4 (LX) complex, which proceeds as the final ligation step in NHEJ [39].

CONCLUSION

A genotoxic agent is a drug or a chemical that can cause aberrations or mutations in the DNA structure and may lead to cancer. They act by altering the chromosomal structures, forming rings, breaks, joins, etc. These can be identified by chromosomal aberration testing. Any drug that prevents the genotoxic effect of clastogenic agents is referred to as an anti-clastogenic or anti-mutagenic agent. The major management of genotoxicity deals with:

- DNA repair methods
- Metabolism of harmful chemical clastogens
- Utilisation of anti-cancer drugs.

The drugs used for the treatment and management of genotoxicity also act as anti-cancer agents, for example, alkylating agents, intercalating agents and enzyme inhibitors. Certain plant extracts such as flavonoids and ellagic acid are found to seize pharmacological activity and are being used as anti-mutagenic agents.

REFERENCES

1. De Flora, S., Izzotti, A. 2007. Mutagenesis and cardiovascular diseases Molecular mechanisms, risk factors, and protective factors. *Mutat Res* 6211:5–17.
2. Natarajan, A.T. 1993. Mechanisms for induction of mutations and chromosome alterations. *Environ Health Perspect* 101:225–29.
3. Silva, R., Sousa, N., Graf, U., et al. 2008. Antigenotoxic effects of *Mandevilla velutina* (Gentianales, Apocynaceae) crude extract on cyclophosphamide induced micronuclei in Swiss mice and urethane induced somatic mutation and recombination in Drosophila melanogaster. *Genet Mol Biol* 31:751–58.
4. Savage, J. 1976. Classification and relationships of induced chromosomal structural changes. *J Med Genet* 13:103–22.
5. Eker, A.P., Quayle, C., Chaves, I., et al. 2009. DNA repair in Mammalia cells: Direct DNA damage reversal: Elegant solutions for nasty problems. *Cell Mol Life Sci* 66:968–80.
6. Phillip, D., Arlt, V. 2009. Genotoxicity: Damage to DNA and its consequences. *EXS* 99:87–110.
7. Sedgwick, B., Bates, P., Paik, J., et al. 2007. Repair of alkylated DNA: Recent advances. *DNA Repair (Amst)* 6:429–42.
8. Goncacak, M., Mircigil, C. 2009. Genotoxicity tests from biomarker studies to the regulations: National perspective. *J Pharm Sci* 34:217–32.
9. Cimino, M. 2006. Comparative overview of current international strategies and guidelines for genetic toxicology testing for regulatory purposes. *Environ Mol Mutagen* 47:362–90.
10. https://www.intechopen.com/books/microwave-materialscharacterization/experimental-requirements-for-in-vitro-studiesaimed-to-evaluate-the-biological-effects-of-radiofreq (accessed March 12, 2020).
11. Shah, S.U. 2012. Importance of genotoxicity & S2a guidelines for genotoxicity testing for pharmaceuticals. *IOSR J Pharm Biol Sci* 1:43–54.
12. Philomena, G. 2011. Concerns regarding the safety and toxicity of medicinal plants -An overview. *J Appl Pharm Sci* 1:40–4.
13. Hanahan, D., Weinberg, R.A. 2000. The hallmarks of cancer. *Cell* 100:57–70.
14. Bushra, M., Mohammed, K., Nahi, Y. 2009. Anti mutagenic effects of *Thymus syriacus* extract. *J Duhok University* 12:216–26.
15. Tuteja, N., Tuteja, R. 2001. Unravelling DNA repair in human: Molecular mechanisms and consequences of repair defect. *Crit Rev Biochem Mol Biol* 36:261–90.
16. Bannister, L., Schimenti, J. 2004. Homologous recombinational repair proteins in mouse meiosis. *Cytogenet Genome Res* 107:191–200.
17. Tweats, D., Scott, A., Westmoreland, C., et al. 2007. Determination of genetic toxicity and potential carcinogenicity *in vitro* challenges post the Seventh Amendment to the European Cosmetics Directive. *Mutagenesis* 22:5–13.

2 Genotoxicity and DNA Repair

INTRODUCTION

All living organisms are continually exposed to endogenous and exogenous agents that interact with the fundamental cellular components. Genotoxicity refers to those agents that interact with DNA or the cellular machinery that regulates the reliability of the genome [1]. A mutation is defined as an eternal alteration in the amount as well as the structure of the genetic material of an organism, which results in a heritable change in the characteristics of the organism involving either single gene, gene blocking or whole chromosomes. DNA damage and mutation are genotoxic endpoints, and the presence of genotoxicity is not indicative of mutation(s). Mutations lead to an eternal alteration in the protein transcribed from the mutated gene, which results in either an abnormal or an inactive protein. If a mutation occurs within a gene and/or the central areas related to their transcription, a dormant gene gets switched on or an active gene gets switched off [2].

Two important gene types are targets for mutations, which lead to neoplasia:

A. **Proto-oncogenes**
   Proto-oncogenes are genes that code for proteins controlling, for example, signalling, growth or the cell cycle. Once they are activated, they transform to an oncogene, which results in eternal over-expression of a protein resulting in cell division. An example is the *ras* gene.

B. **Tumour suppressor** genes
   Tumour suppressor genes code for proteins that inhibit cell proliferation. An example is *p53*.

Gene mutations are alterations in the base-pair sequence of DNA strands. Chromosomal aberrations can be both structural and numerical.

A. **Structural chromosomal aberration** refers to the breaks in chromosomes. These breaks cause chromosomal rearrangements if the ends of broken chromosomes fuse incorrectly.

B. **Numerical chromosomal aberrations** (aneuploidy) are changes in the number of chromosomes per cell, which is caused either by loss or gain of chromosomes during cell division.

Note: In general, chemicals cannot exclusively induce gene mutations or chromosomal aberrations, but they can have a predominant effect [3].

Maintenance of the DNA integrity is essential to guarantee the unrelenting life of the cell and its best possible function according to the given cell type and developmental state. Consequently, the cell develops diverse protective mechanisms to safeguard the integrity of its genetic information and DNA function. These protection mechanisms include:

A. The prevention of DNA damage; anti-oxidant mechanisms to defend against oxidative agents.
B. DNA repair pathways to repair damage after it has occurred.
C. Cell cycle holds to allow the cell time to repair the DNA damage before replication.
D. Apoptosis (programmed cell death) allows multicellular organisms to selectively get rid of excessively damaged cells [4].

However, none of these protective mechanisms is completely effectual, and unrepaired DNA damage can be transformed into a mutation by DNA replication. If a mutation occurs in the non-coding area of the genome apart from the areas regulating transcription, it is unlikely to have any effect on the phenotype of the cell, and mutations in these areas are of fewer corollaries to the overall health of the organism. Mutations within genes are known to accumulate over the lifespan of the cell and/ or organism [5].

DNA DAMAGE

The human genome consists of approximately 3×10^9 bases in a sequence produced from four nitrogenous bases: Adenine (A), Cytosine (C), Guanine (G) and Thymine (T). 5-Methylcytosine (MeC), an analogue of cytosine, has an imperative role in regulating the transcription of genes [6].

DNA is formed of two strands of nitrogenous bases bound via a strict pairing where A binds only to T and C only to G. This pairing is crucial for the DNA replication mechanism, where a fresh strand is produced by "reading" the opposite strand and inserting the correct pairing base. These bases are transformed into proteins using sets of three bases, well-known as a triplet or codon. These codons code for one of the 20 amino acids. Although DNA damage normally affects a single base, it can even more than one base at a time, including the complete chromosome. Cells within the human body are incessantly exposed to DNA detrimental agents, either from endogenous or exogenous sources [7].

The main sources of endogenous DNA damage include:

a. *Hydrolytic damage* – Hydrolytic damage is caused by spontaneous reactions between cellular components and water to produce apurinic sites, that is, where A or G is missing or deamination of A, C or G, which then mispairs.
b. *Oxidative damage* – Oxidative damage is caused by the by-products of intermediary metabolism.

Hydrolytic damage results from spontaneous reactions between cellular components and water to produce apurinic sites, that is, where A or G is missing or deamination of A, C or G, which then mispairs. Oxidative damage results from the interaction between the highly reactive hydroxy radical and DNA to form adducts, such as 8-hydroxyguanine and thymine glycol, and phosphodiesters in the DNA backbone.

DNA damage can arise from exposure to both natural and artificial exogenous agents, including dietary contaminants, natural constituents and chemicals that interact with DNA. In addition, DNA damage and mutations are induced by radiation [8].

The main sources of the exogenous chemical DNA damage are alkylating agents, amines, episulphonium ions, bulky adduct formers, cross-linkers, topoisomerase II inhibitors and oxidising agents. Some of the DNA damaging agents act directly, whereas others act indirectly by metabolic activation reactions, which comprise oxidation and conjugation pathways [9].

Alkylating agents – Alkylating agents (e.g. diethylnitrosamine) transfer an alkyl group, normally methyl or ethyl, to nucleophilic sites on one of the bases. Alkylation prevents correct base pairing and is consequently mutagenic, whereas methylation does not affect base pairing and is not mutagenic *per se*, even though it may result in apurinic sites. Thus, total DNA methylation does not necessarily acquiesce information on the level of mutation [9].

Amines – Aromatic amines, heterocyclic amines and polycyclic aromatic amines react with DNA in the form of a series of reactions starting with oxidation reaction, which generate hydroxylamine and culminate in electrophilic species [9].

Molecules activated by GSH – For several compounds, glutathione conjugation is a detoxification pathway but, for a few chemicals, it results in transformation into a mutagen [9].

Bulky adduct formers – These chemicals bind fully to one of the bases, thereby forming large structures. These large structures do not affect the direct binding of the base pairs, instead they alter the gross structure of the double strand of DNA [9].

Cross-linkers – They either cross-link adjacent bases in the DNA strand (also known as intrastrand) or two opposite bases (also known as interstrand). Intrastrand cross-links are seen by the DNA replication machinery as a single base pair. The interstrand cross-links prevent the two DNA strands from separating to allow the replication and transcription of the DNA strand [9].

Topoisomerase II Inhibitors – Topoisomerase II is an enzyme responsible for cleaving of DNA to facilitate unwinding of the strands for replication or transcription. If inhibited, replication and transcription are obstructed and DNA strand breaks are often formed [9].

DNA REPAIR AND APOPTOSIS

Cells have evolved a range of DNA repair mechanisms to address the various types of damage inflicted on the DNA. Some repair mechanisms are very explicit to distinct types of DNA damage, whereas others detect an extensive range of DNA damage [10–12]. Repair mechanisms fall into five classes:

- base excision repair (BER)
- nucleotide excision repair (NER)
- direct damage reversal
- chromosome break repair and
- mismatch repair

These pathways work together to reduce the unprompted mutation rate to approximately one base-pair change in 10^9–10^{12}.

BER is the undeviating exclusion of a single affected base pair and its replacement with the correct base. The enzymes involved are unambiguous to particular types of damages [10–12]. NER identifies and eliminates an ample spectrum of lesions that cause large and local distortions of the DNA structure, which includes the bulky chemical adducts and some DNA cross-links. The damaged DNA is excised as an oligonucleotide and the resulting slit is filled by DNA polymerases and DNA ligase. BER involves the removal of a single modified base. BER is initiated with the liberation of the amended bases by DNA glycosylases after hydrolytic cleavage. Direct damage reversal is the enzymatic conversion of damaged nucleotides to their original intact state [13].

Two repair systems are responsible for repairing chromosomal breaks:

- Homologous recombination repair (HRR) and
- Non-homologous end joining (NHEJ)

HRR is an error-free system as it restores the chromosomal break based on the homologous sequence from the sister chromatid. NHEJ gives rise to deletions or chromosomal rearrangements as it directly ligates the broken ends at regions of little or no homology. Mismatch repair is, thus, a poorly defined course occurring predominantly post-DNA replication to correct bases that are incorrectly paired or where there are diminutive insertions or deletions in the DNA strand [13].

Once damage to DNA is detected in the cell, signalling proteins delays the cell cycle, which allows time for DNA repair to occur. This signalling is an imperative step to minimise the number of mutations in the cell but can be over-ridden by agents that force the cell into replication and interfere with the signalling process [14–17].

Apoptosis prevents DNA damage from being conceded on to other cells in the entire organism. It is initiated by the cell itself and contrasts with necrosis, where cell death is caused by some damage or infection. Apoptosis is initiated in response to high levels of DNA damage in the cells of multicellular organisms [18–20].

The signal transduction pathways that lead to apoptosis are not fully understood, and it is suggested that probably some unknown genes are involved. One pathway involving *p53* is the pathway involved in the induction of apoptosis followed by DNA damage; it is of principal interest in carcinogenesis. Mutations in proteins involved in apoptosis can be decidedly deleterious as the cell cannot enter apoptosis. Absence of apoptosis has been associated with cancer, as can be established by the connection between mutated *p53* genes and cancer in humans, whereby mutated *p53* genes are present in approximately 60% of cancers, a figure which rises to 90% for skin cancers [21].

DNA repair targets essential areas, such as genome undergoing transcription and DNA strand undergoing transcription. Next to this transcription-coupled repair, the integrity of the genome is preserved regularly by overall genome repair. It should be noted, however, that the majority of this evidence originates from NER research [22].

CONCLUSION

Competent removal of DNA damage from the transcribed sequences leads to the enhancement of cell survival as it enables the cells to express vital genes before the removal of all damaged DNA. Repairing of damaged DNA is targeted within the transcriptionally active strand of the gene, though some does arise on the non-transcribed strand, albeit at a low level. In addition to strand specificity, the repair rate is determined by interplay among:

- Adduct structure
- Accessibility to repair enzymes
- Ability to seize transcription and the DNA conformation.

Note: The assays involved have been discussed in detail in the next chapters.

REFERENCES

1. Adler, I.D. 1984. Cytogenetic tests in mammals. In: Venitt S, Parry JM (eds) *Mutagenicity Testing: A Practical Approach*.s IRL Press, Oxford, Washington, DC. 275–306.
2. Albertini, S., Benthem, J., Corvi, R., et al. 2005. Report on micronucleus test (MNT) in vitro.
3. Ame, B.N., McCan, J., Yamasaki, E. 1975. Methods for detecting carcinogens and mutagens with the Salmonella/mammalian microsome mutagenicity test. *Mutat Res* 31:347–364.
4. Ashby, J., Lefevre, P.A., Burlinson, B., Penman, M.G. 1985. An assessment of the in vitro rat hepatocyte DNA repair assay. *Mutat Res* 156:1–18.
5. Bertram, J.S. 2001. The molecular biology of cancer. *Mol Asp Med* 21:167–223.
6. Chu, E.H.Y., Malling, H.V. 1968. Mammalian cell genetics. II. Chemical induction of specific locus mutations in Chinese hamster cells in vitro. *Proc Nat Acad Sci USA* 61:1306–1312.
7. Cleaver, J. 1984. *Methods for Studying Excision Repair of Eukaryotic DNA Damaged by Physical and Chemical Mutagens. Handbook of Mutagenicity Test Procedures*, 2nd edition, Elsevier Scientific, Amsterdam. 33–70.
8. Cleaver, J., Thomas, G. 1981. *Measurements of Unscheduled DNA Synthesis by Autoradiography. A Laboratory Manual of Research Procedures*. Marcel Dekker, New York. 277–287.
9. Colvin, M., Hatch, F., Felton, J. 1998. Chemical and biological factors affecting mutagenic potency. *Mutat Res* 400:479–492.

10. Dycaico, M., Provost, G., Kretz P., et al. 1994. The use of shuttle vectors for mutation analysis in transgenic mice and rats. *Mutat Res* 307:46–478.

11. Evans, H. 1976. *Cytological Methods for Detecting Chemical Mutagens. Chemical Mutagens, Principles and Methods for their Detection*, Plenum Press, New York and London. 4:1–29.

12. Fenech, M., Morley, A. 1985. Measurement of micronuclei in lymphocytes. *Mutat Res* 147:29–36.

13. Fetterman, B., Kim, B., Margolin, B., et al. 1997. Predicting rodent carcinogenicity from mutagenic potency measured in the Ames Salmonella assay. *Environ Mol Mutagen* 29:312–322.

14. Galloway, S., Armstrong, M., Reuben, S., et al. 1987. Chromosome aberration and sister chromatid exchanges in Chinese hamster ovary cells: Evaluation of 108 chemicals. *Environ Mol Mutagen* 10:1–175.

15. Gossen, J., de Leeuw, W., Molijn, A., et al. 1993. Plasmid rescue from transgenic mouse DNA using LacI repressor protein conjugated to magnetic beads. *Biotechniques* 14:624–629.

16. Gossen, J., de Leeuw, W., Tan, C.H., et al. 1989. Efficient rescue of integrated shuttle vectors from transgenic mice: A model for studying mutations in vivo. *Proc Natl Acad Sci USA* 86:7971–7975.

17. Hahnfeldt, P., Panigraphy, D., Folkman, J., et al. 1999. Tumour development under angiogenic signalling: A dynamical theory of tumour growth, treatment response and post vascular dormancy. *Cancer Res* 59:4770–4775.

18. Heddle, J., Salamone, M., Hite, M., et al. 1983. The induction of micronuclei as a measure of genotoxicity. *Mutat Res* 123:61–118.

19. Kim, B., Park, J., Fournier, D., et al. 2002. New measure of DNA repair in the single-cell gel eletrophoresis (comet) assay. *Environ Mol Mutagen* 40:50–56.

20. Kirkland, D. 1989. *Statistical Evaluation of Mutatgenicity Test Data*. Cambridge University Press, Cambridge, UK.

21. Volders, M., Sofuni, T., Aardema, M., et al. 2003. Report from the in vitro micronucleus assay working group. *Mutat Res* 54:97–100.

22. Kohler, S., Provost, G., Fieck, A., et al. 1991. Spectra of spontaneous and mutagen-induced mutations in the lacI gene in transgenic mice. *Proc Natl Acad Sci USA* 88:7958–7962.

3 Consequence of DNA Damage

INTRODUCTION

DNA mainly contains three cogs:

- Pentose sugar
- Nitrogenous base
- Phosphate group

DNA enables the growth, developmental and reproductive functions in living beings. It is mainly composed of nucleic acid and proteins, lipids and carbohydrates. These macromolecules are vital for living beings. DNA has two strands coiled around each other forming a double-helical structure. Because two strands are formed from monomer units referred to as nucleotides, they are termed as polynucleotides.

A nucleotide is defined as an aggregation of nitrogenous bases such as purines (A, G) and pyrimidines (C, T, U), deoxyribose sugar and phosphate group [1–7]. A number of nucleotide units form the backbone of a DNA helical structure, with the formation of phosphodiester bonds. Two nitrogenous bases of the opposite strand are joined by forming two (A $\equiv$ T) or three (G $=$ C) hydrogen bonds.

A similar sequence is carried by both anti-parallel strands of DNA. Each turn of DNA helix contains ten nucleotide units which are maintained at a distance of 3.4 Å. Therefore, the length of one turn becomes 34 Å. The angle between two nucleotides is 36°.

DNA is wrapped by histone proteins that form a chromatin network.

DNA MUTILATION/DAMAGE

Damage to DNA is an alteration in the primary structure of the double-helical structure. It occurs primarily due to:

- Environmental factors
- Normal metabolic processes inside the cell

This occurs at a rate of 10,000–1,000,000 molecular abrasions per cell per day. The alteration in the bases crops up and can perturb the molecule's customary helical arrangement by forming non-native chemical bonds, which are referred to as bulky adducts [8]. DNA damage can occur due to two main agents:

- Endogenous cellular processes
- Exogenous agents

The endogenous cellular processes comprise:

- Oxidation of nitrogenous bases.
- Generation of DNA strand intrusions from ROS, alkylation of bases, hydrolysis reactions such as deamination, de-purination, de-pyrimidation and bulky adducts formation.
- Mismatch of bases occurs because of errors in the replication process of DNA, mono-adduct dent due to change in mono-nitrogen base and di-adduct damage.

Exogenous agents such as UV radiations (UV-B light) are responsible for

- Direct DNA damage by cross-linking between C and T nitrogenous bases creating pyrimidine dimers, and
- Indirect damage by creating radicals.

Ionising radiations cause damage to DNA by either radioactive decay or cosmic rays, which result in strand breaks. Depurination and single-strand breaks are caused by thermal factors, such as elevated temperatures, which, in turn, affect the DNA helical structure.

Industrial chemicals present in smoke, soot and tar generate various DNA adducts like vinyl chloride, hydrogen peroxide and polycyclic aromatic hydrocarbons, and further lead to the generation of DNA adducts like etheno bases, oxidised bases, alkylated bases, phosphotriesters and cross-linking of DNA.

Nuclear DNA damage occurs inside the nucleus whereas mitochondrial DNA damage occurs inside mitochondria. ROS or free radicals generate a highly oxidative environment and cause *mt*DNA damage. Senescence is an irreversible process by which cell doesn't divide and damaged DNA can't be further replicated. DNA damage is different from mutation. It occurs physically and can be refurbished, but the mutation changes the base sequence that cannot be reversed.

CONSEQUENCES OF DNA DAMAGE

DNA DAMAGE IN MAMMALIAN REPRODUCTIVE CELL

The damage in mammalian germ cells can be roughly repaired in the small time period to provide for the maintenance of genomic heredity [9,10]. The DNA damage in the germ cell is of three types: the damage is repaired immediately, survived and apoptosis.

DNA repair system in reproductive cell repairs genomic mutations caused by genotoxicant factors or addresses a limited number of mutations. The key repair mechanisms in germ cells are NER, BER, MMR, DSBR, etc. DNA damage in male germline cell is related to meagre fertilisation rate which is followed by IVF, substandard pre-implantation, embryonic development and high rate of miscarriage and morbidity in the offspring, which includes childhood cancer. This damage is inadequately differentiated but is known to be involved in the hypomethylation of key genes, oxidative base damage, endonuclease-mediated cleavage and configuration of adducts with xenobiotic and the products of lipid peroxidation.

There are innumerable reasons of DNA damage which are inclusive of futile and ineffective apoptosis, including the oxidative stress associated with male genital tract contagion, exposure towards redox cycling chemicals and imperfection of spermiogenesis connected with the withholding of excess residual cytoplasm.

Physical factors such as exposure to radiofrequency, electromagnetic radiation or gentle scrotal heating can also induce DNA damage in mammalian spermatozoa. However, the fundamental mechanisms are not clear. Ultimately, resolving the specific DNA lesions present in spermatozoa of infertile men will be a significant step towards uncovering the aetiology of this damage and developing approaches for its clinical management.

AGEING – STATE OF DETERIORATION

The decline in homeostasis and the ever-increasing probability of poor health and bereavement which is followed by a progressive organic functional decline is termed as ageing [11,12].

Accrual of DNA damage is a consequence of ageing but not a most important cause. Factors such as diet, lifestyle and exposure to radiation and some genotoxic chemicals give an impression of having a noteworthy effect on the relationship between collective DNA damage and age. Ageing and allied diseases are obsessed with the damaged pathways that drive developmental growth. Nutrient-sensitive signalling network that controls growth drives hyper-functioning of the cell and, thus, reproduction. It is inclusive of insulin, insulin-like growth factor 1 (IGF-1) and meticulous target rapamycin (TOR) kinase pathway.

Several presumptions about ageing have a basis in DNA changes. These presumptions explain age-related transformations in DNA as a corollary of stochastic events. Most important stochastic explanation of DNA damage and other macromolecular smash-up accumulation with age is the oxidative stress theory.

Ageing occurs due to the prolonged activity of pathways and courses during adulthood concerning optimisation of development to this life stage and is, therefore, not due to the accrual of DNA damage. Age-allied diseases are correlated with augmentation of DNA smash-up and mutation or decline in DNA repair.

Deficient DNA repair that results in tissue degeneration and premature ageing is thus indicated by several human genetic defects, such as Cockayne's syndrome (CS) and xeroderma pigmentosum (XP). XP patients have potential skin and eye photosensitivity exhibiting untimely cutaneous ageing, augmented frequency of basal cell carcinoma and melanoma.

LYMPHOIDAL CELLS AND DNA DAMAGE

These are interrelated and allied with planned and involuntary DNA breaks in lymphoid cells. DNA double-strand breaks (DSBs) represent a large number of harmful DNA damage. Such breaks are also produced in a planned manner in mammalian lymphoid cells. B cell receptor and T cell receptor centre consists of variable genes (V) diversity genes (D) and joining genes (J) that recombine in a process known as V(D)J recombination. This is a mechanism that generates a wide repertoire of β cells and T cells and enables these cells to recognise an almost unlimited number of different antigens. The process of this recombination is tightly controlled and DSB must be repaired correctly to avoid persistent deleterious DNA lesions or the resulting translocation. Matured β cells endure two types of genomic alterations:

- Class switch recombination
- Somatic hyper-mutation [13]

This has rigorous and relentless significance for individuals suffering from immune deficiency and neurological blemish to a predisposition to the development of malignancies. Human lymphoid deficiencies include and encompass a broad variety of disarrays with extensive symptoms. The aetiology of these ailments is not completely known.

Genes involved in the non-homologous end, connecting DNA repair pathways, facilitate the repair of planned DNA breaks, which ultimately plays a major role in B cell and T cell disorders. In modern times, there have been numerous reports unfolding how proteins are involved in the NHEJ pathway function as well as their contribution to suppressing the human immune deficiencies and cancer.

NEURODEGENERATIVE DISORDERS

Accretion of DNA abrasions in neurons is associated with neurodegenerative disorders, such as Alzheimer's disease, Huntington's disease and Parkinson's disease. The rationale for this may be that neurons by and large exhibit far above the mitochondrial respiration and couple with ROS production that can smash-up mitochondrial DNA. This damage in the DNA elicits neuronal dysfunction along with degradation. Overall, the nervous system is susceptible to DNA damage because of some degree of capacity of the nervous system for cell replacement in adults. Cells of the nervous system are prospectively differentiated and do not repair by DSB, but HRR must be used for slip-up-level of NHEJ [14].

Neurons significantly rely on transcription and an oxidative DNA damage can wedge this. Thus, the accretion of DNA damage in repair-defective patients possibly leads to ageing. Normal individuals might take away neurons of vital transcripts which leads to cell dysfunction or apoptosis (cell death). Such course of action has a say in the neurodegeneration which is observed in ataxias and CS which is caused by imperfections in DNA strand break repair and transcription couples NER.

GENOME INSTABILITY AND HERITABLE DISEASES

DNA replication causes known diseases because of the growth or retrenchment of genetically unstable DNA repeat sequences, which is usually a trinucleotide motif within a specific locus. For apiece disease, this wavering is thought to arise through the rhythmic nature of these regions allowing damaged DNA secondary structure formation, either during DNA replication or DNA repair process.

Mutations and rescheduling of *mt*DNA can result in impaired mitochondrial function such as amyotrophic lateral sclerosis, mitochondrial encephalomyopathy, Leigh's syndrome, myoclonic epilepsy, Leber's hereditary optic neuropathy and additional myopathies [15–18].

CARDIOVASCULAR DISEASES AND METABOLIC SYNDROME

Pro-apoptic activity 53 (p53)-induced cell death guards against tumourigenesis. This activity is detrimental in settings such as stroke or heart attack. Induction of p53 by either oxidative stress or other sources of DNA damage can also effect the progress of atherosclerosis, thereby providing an association between DNA damage and cardiovascular disease.

Mounting evidence points to human atherosclerosis which is characterised by enhanced DNA smash-up and ultimately leading to senescence of vascular smooth muscle cells. Metabolic syndrome is, thus, a common condition characterised by atypical glucose metabolism and insulin resistance. The malfunction of DNA repair engenders cell proliferation (creation), apoptosis (cell death) and mitochondrial dysfunction. This, in turn, leads to ketosis, hyper-lipidism, amplified fat storage and metabolic syndrome [19–22].

CANCER

Cancer and DNA damage have a close relationship. An elementary feature of this is genome instability. For example, genomic instability in lymphoid tumours frequently corresponds to chromosomal translocation where proto-oncogenes loci are amalgamated to those of antigens receptor apparently by abnormal antigen-receptor recombination. Chromosomal volatility is seen in most infrequent solid tumours. Transient chromosomal wavering likely arises when telomeres in a nascent tumour become decisively short and prone to chromosomal fusion.

Activated oncogenes and the resulting DNA replication stress with DSB formation stimulate chromosomal instability incessantly. At generously proportioned stages of cancer progression, continual hypoxia and cycles of hypoxia and re-oxygenation valour also contribute to genome instability [23–27].

Most carcinogens manoeuvre by DNA damage and thereby lay the grounds for mutations. The elementary cause of cancer is damaged or faulty genes, the directions that enlighten cell what to do. Genes are programmed with DNA; this sets a criterion that anything that damages DNA can augment the risk of cancer. However, a number of genes in the same cell needs to be damaged before it can become cancerous. Most cancer is caused by DNA damage that accrues over a person's lifetime. Some cancers, however, have genetic risk factors.

VIRAL, PARASITIC AND CERTAIN OTHER DISEASES

DNA damage snaps protein function during the lifecycle of human parasites and pathogens, which suggests that DDR inhibitors could be dilapidated to treat similar pathogenesis. For instance, the dependence of HIV on host cell DNA damage response factor recommends a potential for inhibitor in AIDS therapy. The treatment for viral and parasitic diseases needs to be evaluated for probable side effects [28,29]. UV-induced DNA damage have a higher share of radiation bordering between UV-B and UV-A. Such solar UV radiation should create a higher fraction of Dewar isomers. Because the photoisomerisation is largely proficient around 320 nm, it corresponds to the UV absorption maximum of 4-4 PPs.

Accordingly, it has been suggested that all 6-4 PPs should be transformed into Dewar isomers upon exposure to sunlight. Both Cyclobutane Pyrimidine Dimers (CPD) and 6-4 PPs comprise around 75% and 25%, respectively, of UV-induced DNA damaged products. Both can warp the DNA helical structure.

The ability of UV radiation to damage a given base is dogged by the flexibility of DNA. The arrangement of the bases plays a key role since giving out of the dimeric photoproduct highly depends on the pyrimidine bases involved. Sequences that assist binding and unwinding are a favourable site for damage formation [30]. CPD formation is less frequent when there is binding of the DNA towards the minor furrow.

The transcription factor having the direct formation of 6 PPs is in the *TATA BOX* where the DNA is bent, but CPDs are created at the edge of *TATA BOX* and outside where DNA is not bent. CPDs have been reported to be produced in a privileged manner at the major p53 mutational hotspot in UV-B-induced mouse skin tumours.

The natal effect of CPDs has been studied in microbes and mammalians. Thus, they have been reported to hinder the programmes of DNA polymerase. The DNA lesions, if unpaired, may impede with DNA transcription and replication and lead to a misreading of genetic code, thereby causing mutation and death [30].

CONCLUSION

This chapter provides a brief review on the outcomes of DNA damage, such as the effects on the nervous system, cardiovascular system, lymphatic system, as well as various other disorders, such as ageing, cancer and genome instability. This chapter also highlights the diverse innate mechanisms of DNA repair concerning specific DNA damage.

REFERENCES

1. Albert, L., David, L., Michael, M. 2009. Lehringer Principles of Biochemistry. 28–31.
2. Travers, A., Muskhelishvili, G. 2015. DNA structure and function. *FEBSJ* 282:2279–2295.
3. Sinden, R., Pearson, C., Potaman, V., et al. 1998. DNA: Structure and function. *Adv Genome Biol* 5:1–141.
4. Wolffe, A. 1995. DNA structure and function. *FEBS J* 20:330–331.
5. Herbert, A., Rich, A. 1999. Left-handed Z-DNA: Structure and function. *Genetica* 106:1–19.
6. Rich, A., Zhang, S. 2003. Z-DNA: The long road to biological function. *Nat Rev Genet* 4:566–572.
7. Hakem, R. 2008. DNA-damage repair; the good, the bad, and the ugly. *EMBOJ* 274:589–605.

8. Ozturk, S., Demir, N. 2011. DNA repair mechanisms in mammalian germ cells. *Histol Histopathol* 26:505–517.

9. Jackson, S., Bartek, J. 2009. The DNA-damage response in human biology and disease. *Nature* 461:1071–1078.

10. Soares, J., Cortinhas, A., Bento, T., et al. 2014. Aging and DNA damage in humans: A meta-analysis study. *Aging Albany NY* 6:432–439.

11. Menck, C., Munford, V. 2014. DNA repair diseases: What do they tell us about cancer and aging. *Genet Mol Biol* 37:220–233.

12. Prochazkova, J., Loizou, I. 2016. Programmed DNA breaks in lymphoid cells: Repair mechanisms and consequences in human disease. *Immunology* 147:11–20.

13. Jeppesen, D., Bohr, V., Stevnsner, T. 2011. DNA repair deficiency in neurodegeneration. *Prog Neurobiol* 94(2):166–200.

14. Alt, F., Zhang, Y., Meng, F., et al. 2013. Mechanisms of programmed DNA lesions and genomic instability in the immune system. *Cell* 152:417–429.

15. Ghosal, G., Chen, J. 2013. DNA damage tolerance: A double-edged sword guarding the genome. *Transl Cancer Res* 2:107–129.

16. Tubbs, N. 2017. Endogenous DNA damage as a source of genomic instability in cancer. *Cell* 168:644–656.

17. Giuseppe, P., Giovana, B., Ramiro, B., et al. 2015. Metabolic syndrome and DNA damage: The Interplay of Environmental and Lifestyle Factors in the Development of Metabolic Dysfunction. *OJEMD* 5:65–76.

18. Shimizu, I., Yoshida, Y., Suda, M., et al. 2014. DNA damage response and metabolic disease. *Cell Metab* 20:967–977.

19. Mercer, J., Cheng, K., Figg, N., et al. 2010. DNA damage links mitochondrial dysfunction to atherosclerosis and the metabolic syndrome. *Circ Res* 107:1021–1031.

20. Prasad, M., Bronson, S., Warrier, T., et al. 2015. Evaluation of DNA damage in Type 2 diabetes mellitus patients with and without peripheral neuropathy: A study in South Indian population. *J Natl Sci Biol Med* 6:80–84.

21. Lakin, N., Jackson, S. 1999. Regulation of p53 in response to DNA damage. *Oncogene* 18:7644–7655.

22. Lee, S., Chan, J. 2015. Evidence for DNA damage as a biological link between diabetes and cancer. *Chin Med J* 128:1543–1548.

23. Torgovnick, A., Schumacher, B. 2015. DNA repair mechanisms in cancer development and therapy. *Front Genet* 6:157.

24. Gasser, S., Raulet, D. 2006. The DNA damage response, immunity and cancer. *Semin Cancer Biol* 165:344–347.

25. Stoyan, C., Rumena, P., George, R., et al. 2014. DNA repair and carcinogenesis. *Biodiscovery* 12:1.

26. Brégnard, C., Benkirane, M., Laguette, N. 2014. DNA damage repair machinery and HIV escape from innate immune sensing. *Front Microbiol* 5:176.

27. Ellis, L., Robert, H., Roger, J., Grand, R.J. 2016. Activation of the DNA damage response by RNA viruses. *Biomolecules* 6:2.

28. Sinha, R., Häder, D. 2002. UV-induced DNA damage and repair: A review. *Photochem Photobiol Sci* 14:225–236.

29. Ravanat, J., Douki, T., Cadet, J. 2001. Direct and indirect effects of UV radiation on DNA and its components. *J Photochem Photobiol B* 63:88–102.

30. Jacobs, A., Schär, P. 2012. DNA glycosylases: In DNA repair and beyond. *Chromosoma* 121:1–20.

4 Mechanism of Genotoxicity

INTRODUCTION

Chemicals that damage DNA which leads to mutation or cancer are referred to as genotoxic. Toxicological studies have undergone a significant evolution in the previous decade, with much greater emphasis being placed on the following:

- Chronic toxicity
- Carcinogenicity
- Teratogenicity
- Mutagenicity

The mutations in somatic cells are not only implicated in the progression of carcinogenesis but also play a vital role in the pathogenesis of other diverse chronic degenerative diseases, such as atherosclerosis and heart diseases, which are the foremost causes of death among humans [1,2].

Micronucleus test and chromosomal aberration test are used for swot analysis of the anti-mutagenic effect of a drug. One of the best ways to curtail the effect of mutagens and carcinogens is the identification of the anti-clastogens/anti-mutagens, which curb or restrain the process of mutagenesis by acting directly on the mechanism of the cell, as well as desmutagens, which devastate or inactivate, partially or fully, the mutagens, thereby affecting lesser cell population [1,2].

Nature has given us medicinal plants which need to be explored for their use as either anti-mutagenic and anti-carcinogenic food or drug additives [1,2].

According to genetics, genotoxicity describes the varied chemical means that damage the genetic information within a cell causing mutations, which may lead to cancer. While genotoxicity is frequently confused with mutagenicity, it is imperative to note that all mutagens are categorised as genotoxic; however, not all genotoxic substances are categorised as mutagenic.

The alterations can have undeviating or circuitous effects on DNA

- The initiation of mutations,
- Mistimed event activation, and
- Direct DNA damage leading to mutations.

The permanent, hereditary changes can affect either somatic cells or germ cells, which pass it on to future generations. Cells thwart expression of the genotoxic mutation by either repairing of DNA or apoptosis (cell death); however, the damage may not always be fixed, which ultimately leads to mutagenesis [1,2].

To assay for genotoxic molecules, researchers need to assay for DNA damage in cells exposed to noxious substrates. This DNA damage can be in the form of:

- Single- and double-strand breaks
- Loss of expurgation repair
- Cross-linking
- Alkali labile sites
- Point mutations
- Structural and numerical chromosomal aberrations

The compromised reliability of the genetic material has been known to form the basis of cancer. Consequently, many sophisticated methods/techniques, including Ames Assay, *in-vitro* and *in-vivo* toxicology tests and Comet assays have been urbanized to gauge the chemicals' probability to cause DNA damage, which may, in turn, lead to cancer [2]. ·

Anti-mutagens have been described as agents that reduce the perceptible yield of impulsive and/or induced mutations. Mechanism of anti-mutagenesis has been classified into two major processes:

- Desmutagenesis; wherein factors act directly on mutagens or inactivate them.
- Bioantimutagenesis; wherein factors act on the course of mutagenesis or repair DNA damages that effect the mutation frequency.

Gemcitabine is used as a mutagen with an anti-metabolite activity; it acts by prohibiting elongation of DNA chain. Anti-mutagenesis is considered one of the most viable ways for inhibiting the off-putting effects of environmental genotoxicants inclusive of carcinogens. Currently, a large number of anti-mutagens of plant origins are known. Evaluation of genetic toxicity is an imperative module for the safety appraisal of chemicals, including pharmaceuticals, agricultural chemicals, food, additives and industrial chemicals. Currently, genotoxicity has been regulated primarily based on the qualitative results of hazard identification assays, that is, decisions are based on classification as positive or negative for genotoxic potential. Most human carcinogens are recognised by epidemiological studies. These studies are necessarily long term as no effect is expected to be observed until decades after the carcinogenic event or events [3].

However persuasive, these studies are costly and exposure levels and effects are complicated to quantify. A few manifold generation mutation assays have been approved using rodents:

- Dominant lethal
- Mouse spot test
- Heritable translocation test

These tests must be conceded on a large scale and must be insensitive; to perceive a 1% increase (which is a very burly effect) in carcinogenicity in a human population, one would need to conduct an animal study to such a large scale as to cost over a couple of million dollars.

Genotoxicity tests can be defined as *in-vitro* and *in-vivo* tests intended to discover compounds that induce genetic damage by assorted mechanisms. These tests facilitate hazard identification concerning DNA damage and its fixation. Fixing DNA damage in a variety of gene mutations open scale chromosomal damage or recombination is normally considered to be indispensable for heritable effects and in the multi-step growth process known as malignancy, which is a complex process where genetic changes play a partial role [4].

Numerical chromosomal changes have also been associated with tumourigenesis and can point towards a potential for aneuploidy in germ cells. Compounds that are positive in tests and can detect such kinds of damage have the potential to be human carcinogens and/or mutagens. As the involvement between exposure to particular chemicals and carcinogenesis is expected for humans, whereas an analogous relationship has been complex and convoluted to prove for heritable diseases. Genotoxicity tests have, therefore, been used principally for the prophecy of carcinogenicity [5].

However, as germline mutations are allied with human disease, a compound with heritable effects is considered to be just as stern as the suspicion that a compound might induce cancer. In addition, the effect of genotoxicity tests can be important to interpret carcinogenicity studies.

Mutations are variation in the DNA sequence of a cell's genome and originate through radiation, viruses, transposons and mutagenic chemicals, as well as slip-ups that occur during meiosis or DNA replication [5]. There is no consensus among genetic toxicologists regarding the classification of mutations. Three groups of mutations have been distinguished:

- **Single point mutations or Gene mutations**: These are minor changes in the DNA at the base and gene level, which are not visible under a light microscope:
  a. Base pair substitutions
  b. Addition or deletion of bases
- Structural chromosomal aberrations
- Genome mutations

The interaction of the genotoxic substance with the DNA structure and sequence leads to damage to the genetic material. Their interaction is at a specific position, location or base sequence of the DNA structure, which causes lesions, breakage, fusion, deletion, missegregation or non-disjunction leading to injury and mutation [6]. For example, the transition metal chromium in its high-valent oxidation state interacts with DNA, leading to carcinogenesis. It has been evident that the mechanism of the damage caused and base oxidation products for the interface between DNA and high-valent chromium are pertinent and applicable to *in-vivo* formation of DNA damage, which can cause cancer in chromate-exposed population, which marks high-valent chromium as carcinogenic [7].

ROS cause most abundant oxidative lesions in DNA. Oxidants and free radicals present in the cellular system adversely affect and alter the structure of lipids, proteins and DNA. Reactive aldehydes like 4-hydroxynonenal (4-HNE) which are generated by the decomposition of lipid peroxyl radicals or by a primary free-radical intermediate of lipid peroxidation also cause lesions.

4-Hydroxynonenal is involved in many oxidative stress-related diseases, namely, atherosclerosis, fibrosis and neurodegenerative diseases. It can stimulate cell proliferation, differentiation as well as a cytoprotective response through its effects on various signalling pathway [8].

STANDARD TEST BATTERY FOR GENOTOXICITY

The standard tests (assays regarding accounting for genotoxicity) have been summarized in Table 4.1.

TESTING FOR GENE MUTATION IN BACTERIA

- *In vitro*: It is used for the cytogenetic evaluation of chromosomal damage with mammalian cells or mouse lymphoma assay.
- *In vivo*: It is used as a test for chromosomal damage using rodent haematopoietic cells.

TABLE 4.1
Standard Test Battery for Genotoxicity

TG 471	Bacterial Reverse Mutation Test
TG 472	*Escherichia coli*, reverse assay
TG 473	*In-Vitro* Mammalian Chromosome Aberration Test
TG 474	Mammalian Erythrocyte Micronucleus Test
TG 475	Mammalian Bone Marrow Chromosome Aberration Test
TG 476	*In-Vitro* Mammalian Cell Gene Mutation Test
TG 477	Sex-linked Recessive Lethal Test in Drosophila melanogaster
TG 478	Rodent Dominant Lethal Test
TG 479	*In-Vitro* Sister Chromatid Exchange Assay in Mammalian Cells
TG 480	*Saccharomyces cerevisiae*, Gene Mutation Assay
TG 481	*Saccharomyces cerevisiae*, Mitotic Recombination Assay
TG 482	DNA Damage and Repair, Unscheduled DNA Synthesis in Mammalian Cells *In Vitro*
TG 483	Mammalian Spermatogonial Chromosome Aberration Test
TG 484	Mouse Spot Test
TG 485	Mouse Heritable Translocation Assay
TG 486	Unscheduled DNA Synthesis Test with Mouse Liver Cells *In Vitro*
TG 487	*In-Vitro* Mammalian Cell Micronucleus Test

In-vitro Testing Methods

There are various *in-vitro* genotoxicity testing methods available. Some of the commonly used tests have been described in this book, which are also a part of the standard battery as mentioned above. Bacterial reverse mutation test, otherwise known as Ames test, whose endpoint is the gene mutations in the bacterial cell [9].

- Mammalian chromosome aberration test with the endpoint of chromosomal aberration [10–12].
- Mammalian cell gene mutation test or the mouse lymphoma test with the endpoint of gene mutations.

CONCLUSION

Genotoxins interact with the DNA, causing mutations and damage to the structure which leads to cancer. They alter the chromosomal structure either by addition, deletion, duplication or forming rings. These mutations lead to various diseases including cancer.

Therefore, it is important to consider genotoxicity studies to avoid the potential damage caused by genotoxins.

Genotoxicity tests are performed to identify whether a drug or other substances have the potential to cause mutation and genotoxicity. This helps in the identification of the hazards during the early stages of drug development. In addition, it helps in understanding the mechanism of the mutation and genotoxicity, thereby creating improved and enhanced techniques and approaches to both avert and avoid the frequency of mutation and genotoxicity.

REFERENCES

1. Yamada, M., Espinosa, A., Watanabe, M., et al. 1997. Targeted disruption of the gene encoding the classical nitroreductase enzyme in *Salmonella typhimurium* Ames test strains TA1535 and TA1538. *Mutat Res* 375:9–17.
2. Clements, J. 2000. The mouse lymphoma assay. *Mutat Res* 455:97–110.
3. http://www.mindfully.org/Pesticide/Biological-Evaluation-Med-Devices-3.html (accessed March 12, 2020).
4. Mortelmans, K., Zeiger, E. 2000. The Ames Salmonella / microsome mutagenicity assay. *Mutat Res* 455:29–60.
5. Evans, H. 1976. Cytological methods for detecting chemical mutagens, principles and methods for their detection. *Chemicalmutagens* 4:1–29.
6. Maron, D., Ames, B. 1983. Revised methods for the Salmonella mutagenicity test. *Mutat Res* 113:173–215.
7. Ames, B., Mccann, J., Yamasaki, E. 1975. Methods for detecting carcinogens and mutagens with the Salmonella / Mammalian microsome mutagenicity test. *Mutat Res* 31:347–64.
8. Galloway, S., Armstrong, M., Reuben, C., et al. 1978. Chromosome aberration and sister chromatic exchanges in Chinese hamster ovary cells: Evaluation of 108 chemicals. *Environ Mol Mutagen* 10:1–175.
9. Ishidate, M., Sofuni, T. 1985. The *in vitro* chromosomal aberration test using Chinese Hamster Lung (CHL) fibroblast cells in culture. *Mutat Res* 5:427–432.

10. Aaron, C., Bolcsfoldi, G., Glatt, H., et al. 1994. Mammalian cell gene mutation assays working group report. Report of the International Workshop on Standardisation of Genotoxicity Test Procedures. *Mutat Res* 312:235–239.
11. Aaron, C., Stankowski, L. 1989. Comparison of the AS52 / XPRT and the CHO / HPRT assays: Evaluation of six drug candidates. *Mutation Res* 223:121–128.
12. Morita, T., Nagaki, T., Fukuda, I., et al. 1992. Clastogenicity of low pH to various cultured mammalian cells. *Mutat Res* 268:297–305.

5 Impact of ICH Guidelines on Genotoxicity Testing Dogmatic Obligation for Innovation Drug Discovery and Development

INTRODUCTION

Genotoxicity testing of new chemical entities (NCE) is usually performed for recognising the risks of DNA damage and their fixation [1]. This security and compensation can be manifested in the form of gene mutation, structural chromosomal aberration or recombination and numerical changes. These alterations have likely heritable effects on germ cells and thus impose risks to future generations [1]. In addition, it has been well documented that somatic mutations can also play an important role in malignancy [1]. These tests have been used mainly for the prediction of carcinogenicity and genotoxicity because compounds, which are positive in these tests, have the potential to act as human carcinogens and/or mutagens.

BUDGE IN DRUG DISCOVERY HYPOTHESIS

At present, new drug discovery requires a methodical exploration of the drug's safety and effectuality before their release into the market [1]. The financial investment in a very novel drug discovery grows exponentially as the compound progresses from early discovery through development to registration. Recently, the hypothesis of drug discovery methods has been re-worked because of technological developments.

A large variety of diverse molecules are speedily synthesised by combinatorial chemistry, and high-output screening has exaggerated the scope and speed of biological assays for safety analysis [5]. The chief objective of drug safety analysis is to get a biological sequence that is indicative of toxicity, which might be taken and extended to the appraisal of health risk to humans [1]. The potential risks and advantages of the drug under swot are rigorously considered, hence, the compensation of employing a new drug as a therapeutic agent outweighs its risks [1]. New drug discovery may be a multifarious process and NCE moves through completely different phases of development either in a serial manner or in synchronic approach and eventually reaches the therapeutic target.

Toxicity assessment of a brand new molecule is performed in parallel with the pharmacological assessment early during the analysis. Toxicity information provided early during the developmental stages can considerably reduce monetary expenditure through shunning of additional activities and improves potency. Pre-clinical studies are the prime source of data relating to the biological effects of a substance. These results influence not only the choice whether or not to expose human subjects to NCE but also the planning of their clinical trials. The detection of biological adverse effects might lead to the cancellation of the event of the drug or a return to basic chemistry to switch the structure and attenuate the chance.

RESPONSIBILITY OF NATIONAL AND INTERNATIONAL DOGMATIC AUTHORITIES

Most countries have precise guidelines for testing pharmaceuticals for genotoxicity. In India, the Schedule Y of drugs and Cosmetic Rules 1988, Central Drugs Standard Control Organization (CDSCO), Directorate General of Health Services (DGHS), New Medicine Division issued by Ministry of Health and Family Welfare, Govt. of India deal with the stipulations to conduct the clinical trials of a new drug before its promotion, relying upon the standing of the drugs in different countries.

As per the regulative requirements, the mutagenicity and carcinogenicity testing are needed once the compound or its substance is structurally associated with an identified carcinogen or when the character and action of the drugs recommend a mutagenic/carcinogenic potential. In the United States, the Food and Drug Administration's (FBA's) Centre for drugs and Biologics Evaluation and Research (CDER and CBER) suggest genotoxicity testing for all new drugs [1].

At present, the FDA accepts the three-test package *pro re nata* by the Ministry of Health and Welfare (MHW) in Japan. In the European Union (EU), mutagenicity information is needed for prescription drugs before commencing clinical trials [1].

However, further checks can be needed in specific circumstances [1]. In Japan, the MHW adopted mutagenicity tests in 1984 jointly of the several toxicity studies needed for the approval to manufacture or import the latest drugs [2].

The guidelines embrace three representative tests: a reverse mutation assay in bacteria, body aberrations test with mammalian cells in culture and a rodent micronucleus test. These were amended in 1989 after several revisions. The Canadian authorities consider prescription drugs because of the greatest level of initial concern for pre-clinical toxicity testing in view of high and widespread use for humans. The Health Protection Branch (HPB) of the Department of Health and Welfare in Canada needs genotoxicity testing of all drugs based mostly upon the level of concern strategy [2].

The Organization for Economic Cooperation and Development (OECD) guidelines are specifically necessary as compliance in one country can ensure the acceptance of toxicity information in each of the thirty member countries of the OECD [3]. The International Conference on Harmonization (ICH) brings together the regulative authorities Europe, Japan and the United States. When the ICH guidelines were developed, the OECD updated many of their genotoxicity test guidelines. In fact, both ICH and OECD influenced one another leading to an identical recommendation.

However, because of the particular nature of prescription drugs, the ICH guidelines contain pharmaceutical recommendations that considerably differ from the recent OECD guidelines [3].

PRECINCTS OF THE REGULATORY SYSTEM TO TEST GENOTOXICITY

At a glance, it is inconceivable to register a brand new drug without providing information regarding its mutagenicity [4]. The present variations in protocol style and practices between different regulative authorities hinder the drug discovery method and delay the promotion of the potential candidates. As per the regulative necessities to introduce an NCE as a drug, the diagnosing safety study must be performed in every country as per that country's guidelines. These are very time-intense, resource-intensive processes and require various animals for experimentation.

In most countries, the rules are inadequate to draw a definitive conclusion for the genotoxic potential of an NCE. The *in-vitro* and *in-vivo* protocols mentioned within the guidelines don't seem to be valid for the detection of ordination mutations (aneuploidy) and somatic point mutations. For sure categories of pharmaceuticals, which require vital experimental analysis, there aren't any details for the selection of specific test system and test protocols [4].

Most guidelines are empty recommendations for compounds that are genotoxic but seem to act by non-DNA targets. There are no specific recommendations for various genotoxic and tumourigenic compounds and their organ-specific effects once they are used therapeutically [4].

GENOTOXICITY TESTING PROCEDURES USED IN REGULATORY TOXICITY

Over the past few years, numerous countries have issued many tips on the genotoxicity testing of recent drugs. Though no specific tests are prescribed in some countries, it is apparent that almost all the rules suggested by numerous regulative authorities are designed within the four-test battery. These tests include:

- A genetic mutation in a bacterium
- A check for body aberrations in class cells *in vitro*
- A test for genetic mutation in eukaryotic cells *in vitro*
- *In-vivo* test for genetic injury

The compound under investigation should comply with genotoxicity testing once the results of all the tests within the four-test battery are uniformly negative. More experiments are needed if the test results are not uniform within the four-test battery [4–7].

In India, the eighth amendment of the Drug and Cosmetic Rules proposes and counsels for mutagenicity and carcinogenicity testing, which is under Schedule Y before clinical trials intended for the import and manufacture of a new drug. No specific check has been suggested for this purpose. However, a minimum of a three-dose-level mutagenicity check should be applied once the drug or its substance is expounded as a known carcinogen [4–7].

Recently, the Government of India constituted a committee consisting of consultants to re-evaluate the guidelines. The main objective of the committee is to advise the Drug Controller General of India in matters relating regulative provisions under Schedule Y of the drugs, the Cosmetics Rules any modifications needed thereof. The committee will also undertake an in-depth analysis of knowledge for the approval of phase I clinical trial for an Investigational New Drug or NCE.

The schedule of genotoxicity testing about clinical trials is different in numerous regions. Typically, the *in-vitro* mutation check and chromosomal aberration tests are completed before the primary human exposure, and the standard battery needs to be completed before initiating the phase II clinical trials [4–7].

With the implementation of various tips by the regulative authorities of various countries, good laboratory practices are obligatory to evaluate genotoxicity [19–20]. With this, significant reproducible and reliable results can be expected. However, in a majority of the cases, inconclusive genotoxic activity is obtained because of inadequate testing.

The range and type of genotoxicity studies used for prescription drugs don't seem to apply to the active elements of biotechnology-derived pharmaceutical products. Wherever there's a cause for concern about the product, various relevant models need to be developed and the product ought to be tested [4–7]. The results of genotoxicity are required for the planning and conduct of future carcinogenicity studies and the interpretation of the test results [4–7].

PRINCIPLE OF GENOTOXICITY TESTING

Chemicals that exert their adverse effect through interaction with the genetic material (DNA) of cells are referred to as genotoxic [4–7]. Most human carcinogens are genotoxic. The science of genotoxicity mainly considers that chemicals inducing mutations in varied experimental models could conceivably affect the incidence of inheritable mutations in humans [7].

Genotoxicity tests are often outlined as *in-vitro* or *in-vivo* tests designed to notice medication, which may induce genetic injury directly or indirectly by varied mechanisms of action. Genotoxicity tests alter hazard identification concerning DNA injury and its fixation within the mutations, large-scale body injuries, recombination and numerical body changes [7].

Drugs that are positive in these tests have the potential to be human carcinogens and/or mutagens, that is, they could induce cancer and/or inheritable defects.

THE CONCEPT OF TEST BATTERIES

During the last 20 years, many test systems with completely different endpoints have been successfully introduced in routine genotoxicity testing [8]. Three levels (gene, body and genome) of knowledge are needed to provide excellent coverage of a brand new drug substance and its potential [29].

At present, genotoxicity tests seldom observe more than one endpoint in an exceedingly single assay system [30]. Hence, to attenuate the chance factors, an NCE needs to be subjected to a battery of genotoxicity tests. Genotoxic studies in conjunction with pharmacokinetic (PK) and toxicokinetic (TK) information represent a rational approach for the assessment of mutagens and carcinogens [8]. Such data does not solely permit the correlation of nephrotoxic signs and symptoms with blood tissue levels of the agent; however, they conjointly permits the choice of appropriate animal species.

A well-considered look at choice is essential to see the chance potential of NCE properly. Therefore, the same approach is to perform a battery of *in-vitro* and *in-vivo* tests as a part of the toxicologic analysis method. However, there is no universal agreement on the most effective combination of tests for a selected purpose. With the continued refinement of the global need for promoting new molecules by ICH, there has been considerable support and encouragement for obtaining multidisciplinary information for a higher interpretation of the drug safety information [9].

CRITERIA FOR TEST BATTERY SELECTION

For employing a battery of tests in genotoxicity testing, it has continuously been ascertained that *in-vitro* tests play a significant role because of their high sensitivity and rate. Preliminary tests are designed to identify the majority of the genotoxic carcinogens [9].

One of the foremost vital criteria for *in-vitro* analysis is the relative sensitivity of various cell lines and their genetic diversity [9]. For assay duplicability, an additional comprehensive protocol with a clear understanding of vital processes (pH shift, high osmolality, high ionic strength) for analysing throughout assay performance is necessary [9]. The *in-vivo* check models are usually designed to ascertain the chemical effects on the route of exposure, the period of treatment, metabolism and organ exposure which is therapeutically relevant to humans.

DRAWBACKS OF CURRENT GENOTOXICITY TESTS

The majority of the presently used genotoxicity assays for dogmatic toxicity testing were developed in the 1970s. Thus, their throughput cannot meet the current drug discovery requirements [9]. In most cases, the site and mechanism by which genotoxicity is produced by the compound under investigation are unknown.

The target site in the test system might not be the identical target site of toxic action of the NCE. In sub-chronic and chronic toxicity testing, several relevant parameters or endpoints can be perceived for the determination of the toxicity, but the same is hardly ever true for genotoxicity tests.

A single test system cannot be applied for the universal detection of all the germane genotoxic substances. Testing requirements depend upon the nature and category of chemical substances. There is a lack of a validated test system for the detection of induced genome mutation (aneuploidy) in germ cells [9].

ICH GUIDELINES

The pre-clinical testing guidelines have been summarised for genotoxicity as well as other facets of toxicity testing. It fulfils the primary harmonisation targets for the three ICH regions. An ordered approach has been defined from the viewpoint of regulation using a restricted number of well-defined tests that harmonise each other in terms of endpoints and which consent a systematic judgement of genotoxicity as it is necessary.

Recently, an integrated approach has been developed which is inclusive of representatives from the pharmaceutical industry, scientists and regulatory authorities to develop homogeneous guidelines for toxicological testing [10].

Two ICH guidelines have been important in genotoxicity testing. The first genotoxicity working group recommended the following guidelines:

- *S2A: Genotoxicity* – it is the Guidance on Specific Aspects of Regulatory Genotoxicity Tests for Pharmaceuticals.
- *S2B: Genotoxicity* – It is the guidance for a standard test battery for Genotoxicity Testing of Pharmaceuticals.

The S2A guidelines swathe the tactical issues and protocol design for *in-vitro* and *in-vivo* genotoxicity testing.

IN-VITRO STUDIES

- The recognition of mutagens, which induce DNA base changes, is improved by including strains that notices A-T base pair changes.
- Chemicals tested at highly toxic doses *in vitro* can lead to "false positive" results. Therefore, it has been recommended that the test compound should be tested up to a concentration, which creates at least 50% inhibition in cell propagation or mitotic index for cytogenetic tests.
- Some genotoxins are only noticeable when tested in the insoluble range [10]. For the evaluation of the genotoxicity of a compound without considering cytotoxicity, the largest concentration of the test compound to be used is 5 mg/plate and 10 mM for bacterial and mammalian cell culture, respectively. In case of precipitation of the compound, studies ought to be carried out up to the least concentration at which precipitation occurs.

The guideline also supervises the understanding of test results from *in-vitro* tests. In particular, a proposition has been made on bewildering factors that can lead to erroneous interpretation of the data.

IN-VIVO STUDIES

- Both rats and mice are convincing for *in-vivo* analysis of genotoxins.
- As per the test protocol, bone marrow micronucleus test and metaphase investigation are interchangeable for regulatory rationale and only males are employed for testing genotoxins.
- In addition to the above two tests, the micronucleus test in peripheral blood of mice is also adequate for generating reproducible data.

- When a positive response is not obtained, it is obligatory to corroborate the assay by demonstrating exposure to the target tissue. This is predominantly imperative when a "positive result" has been obtained in *in-vitro* genotoxicity tests.

The S2B guideline covers the recognition of a standard set of tests to be conducted for listing, as well as the scope of confirmatory experimentation in *in-vitro* genotoxicity tests in the standard test battery.

The ICH has recommended three standard test batteries for the evaluation of all pharmaceutical products [11]:

- A trial for gene alteration in bacteria.
- An *in-vitro* test with an evaluation of chromosomal alteration in mammalian cells or an *in-vitro* mouse lymphoma TK assay.
- An *in-vivo* test for chromosomal alteration using rodent haemopoietic cells.

It is largely apposite to test genotoxicity in a bacterial reverse mutation (Ames test) assay to get introductory information. This test has been shown to detect base substitutions, frameshift point mutations as well as DNA cross-linking agents [12].

The preponderance of genotoxic carcinogens detected by this test is primarily because of the comparative sensitivity of the test strains to the diverse test compounds and the surveillance of mutation, which occurs at a very low concentration. The *in-vitro* tests employing different mammalian cells play a chief role because they are highly receptive, and widely held genotoxic compounds can be detected.

For deprived bioavailable compounds, achieving target organ exposure in mammalian bone marrow cells is complex. This can be overcome by directly exposing the cells to chemicals in an *in-vitro* system. However, an *in-vivo* test battery is usually taken into consideration as a part of the basic test battery to afford additional relevant factors, such as absorption, distribution, metabolism and excretion that may influence the genotoxic activity of a compound [13].

Therefore, by testing a compound in the three-test battery, as recommended by the ICH, both gene and chromosomal level in sequence can be obtained.

VARIATION OF THE TEST BATTERY

There are certain situations where the standard three-test battery may need additional amendment [14]:

- *Highly toxic compounds* – Compounds that are highly toxic to bacteria and impede the mammalian test system(s) should be tested in two *in-vitro* mammalian cell tests using two different cell types. The endpoint should include the fortitude of gene mutation and chromosomal alteration.
- *Compounds bearing structural alerts* – Compounds that give negative results in the standard genotoxic test battery but have some structural alerts need to be subjected to further additional tests with modified protocols.

- *Compounds that are not absorbed* – Compounds that are not systemically absorbed and, therefore, are not accessible to the target tissue, either bone marrow or liver, should be solitarily tested in *in-vitro* assays. These tests should comprise a bacterial gene mutation and two *in-vitro* mammalian cell assays using two different cell types with two different endpoints.
- *Evidence of tumour response* – Compounds that are negative in standard genotoxic test battery but display carcinogenic potential should be additionally tested in appropriate models to evaluate the mechanism of action associated with the carcinogenic activity. Additional testing includes exogenous metabolic activation or can include:
  a. The liver unplanned DNA synthesis test
  b. ^{32}P post-labelling
  c. Induction of mutations in transgenes
  d. Molecular categorisation of genetic alteration in tumour related genes
- *Structurally unique chemical classes* – On exceptional occasions, an entirely novel compound in a unique chemical class will be introduced as a pharmaceutical. Such compounds that are not subjected to constant rodent carcinogenicity bioassays possibly will be tested further by employing genotoxicity testing using supplementary *in-vitro* and/or *in-vivo* assays.

SPECIFIC RECOMMENDATIONS

In-Vitro Test Results

The *in-vitro* gene mutation tests in both bacterial and mammalian system should comprise range-finding results, which will direct the selection of suitable concentration to be used in the ultimate mutagenicity test. The range pronouncement tests in bacteria are achieved and executed upon on all strains with and without metabolic commencement and instigation with appropriate and apposite positive and negative controls [15].

In the ICH recommendation, the Mouse Lymphoma TK Assay (MLA) is considered to be superior compared to the test using hgprt locus test in mice. In MLA, for short exposure, the test should be inclusive of both with and without metabolic activation with fitting positive and negative controls [15].

In case of negative results without metabolic activation, an incessant exposure for approximately 24 h is obligatory. A negative *in-vitro* result may be additionally tested with the precise external metabolising system, which is known to be proficient for the metabolism/activation of that explicit class of compound under test. The *in-vitro* chromosomal smash-up should be performed as per the protocol mentioned in *in-vitro* gene mutation test, with a supplementary sequence of polyploidy and mitotic index induced by the chemical under investigation [15].

There are several circumstances that can pilot to positive *in-vitro* test results and should be evaluated for their natural relevance. The augmentation in response over negative control should point towards a meaningful genotoxic effect to the target cells. Additional exploration should be performed to exhibit that the *in-vitro* positive results are not because of the extreme conditions (e.g. excess pH, osmolality, heavy precipitate in suspension culture). Comprehensible evidence needs to be recognised that the *in-vitro* positive result did not instigate due to contaminant or a metabolite from *in-vitro* specific metabolic pathways. Thus, the expert working group has described a series of conditions from which positive results of questionable biological relevance might be obtained in mammalian cells *in vitro* [16].

In-Vivo Test Results

The chromosomal aberration assay in rodent bone marrow nucleated cells can perceive an extensive gamut of changes, which results from the breakage of one or more chromatids as the initial event. Breakage of chromatids or chromosomes can affect micronucleus formation if an acentric fragment is produced. Therefore, assay perceiving either chromosomal aberration or micronuclei provides an alert with satisfactory, agreeable and adequate information for the detection of clastogens. Therefore, the extent of micronucleated erythrocytes in polychromatic erythrocytes is a suitable substitute to identify clastogens/aneuploidy inducers. Male mice are more responsive than female mice for induction of micronuclei; however, the disparity is only quantitative, not qualitative [16,17].

DETECTION OF GERM CELL MUTAGENS

Concerning germ cell mutagens, no specific test has been advocated by ICH. It is well documented that mutations in the germline cells mainly lead to the indecent function of germ cells and early embryo loss. Among the abnormalities, aneuploidy is the most common, which is followed by polyploidy. Structural abnormalities comprise about 5% of the total germ cell disorders.

There may be specific types of mutagens, for example, aneuploidy inducers, which precede preferentially during meiotic gametogenesis stages. When assessing the likelihood of a genotoxic potential of a germ cell mutagen, the brand of active compound, pharmacokinetic data and indication for medical use are taken into consideration. This could direct to contemplation that a genetic hazard for human germ cells is not ordinary [18,19].

ADDITIONAL REQUIREMENTS FOR SPECIFIC SITUATION

Male mice bone marrow micronucleus test is sufficient for *in-vivo* genotoxicity testing. In some specific cases, female mice, as well as rats, can be used to ensure the legitimacy of the genotoxicity testing. Reports point out that the metabolism of the compounds may diverge not only within the sexes of the same species but also across the species. No detailed argument has been made under ICH guidelines for *in-vivo* gene mutation assay which employs the transgenic mice model [20].

Some specific mutagens, for example, aneuploidy inducers, act in a privileged manner during meiotic gametogenesis stages, and there is no irrefutable experimental method existing till date to detect these substances. As numerous chemicals produce positive results in *in-vitro* but not in *in-vivo* test system, they are irrelevant in analysing human exposure. Therefore, there is a need for mechanistic studies to determine whether the aberration induced *in vitro* are due to a procedure with or without a brink.

IMPACT OF ICH GUIDELINES

The harmonisation process will augment the haste of the existing regulatory process for NCE. It will also eradicate and abolish the negative aspects of the modern toxicological evaluation of new drugs. This will eventually perk up the accuracy of genotoxicity detection and help to conserve resources. With the foreword of ICH, it has been recommended that all genotoxicity tests should be carried out untimely in the development of pharmaceuticals and should be concluded before the initiation of phase II clinical trials [21].

The guidelines intend to be flexible for a novel group of pharmaceuticals, especially biotechnology-derived products and permit the substitution of a newer generation of validated bioassays.

Once the ICH guidelines are implemented, the process of endorsement by the regulatory authorities will be sleek and will reduce the time and conserve the resources in drug discovery and development process [22].

NOVEL METHODS IN GENOTOXICITY TESTING

For the evaluation and assessment of genetic toxicity of NCE, three standard genetic toxicity test battery is sufficient. However, rarely, the standard battery may be inadequate, which might necessitate further testing. Such supplementary testing may be mechanistic in sequence for chronic carcinogenicity bioassay.

The ICH guidelines do not exclude innovative methods and encourages the development of new systems and their use when sufficient scientific justifications support the findings. Many mutagens form adducts with DNA either directly or indirectly after metabolic activation. Therefore, highly receptive and specific analytical methods like 32P post-labelling [23,24], immunological assays using polyclonal and monoclonal antisera and mass spectrometry are used for adduct analysis [24].

For analysing single and double DNA strand breaks, the Comet assay is a hasty visual method for quantitative estimation [24].

The transgenic mice model provides an opportunity to study *in-vivo* gene mutation and to appreciate the multipart mechanism of carcinogenesis, which has a greater potential for genotoxicity testing. Other tests like consideration of p53 gene mutation, recognition of apoptosis, revealing of aneuploidy by anti-centromere antibody, utilisation of fluorescent in situ hybridisation to visualise translocation of chromosomes, detection of unscheduled DNA synthesis and cell transformation assay can be used for genotoxicity screening. These tests increase both the sensitivity and specificity of the existing test protocols [24].

CONCLUSION

Currently, drug discovery and development is speedy, time profitable and dynamic because of the use of pioneering technologies, such as genomics, high-throughput screening and proteomics. Pharmaceutical companies and other regulatory agencies need to assemble the challenges of the 21st century, and toxicologists and other scientists need to re-evaluate the current protocols in this changing environment.

The accessible guidelines, in a developing country like India, need to be re-evaluated and modified in accordance with the new-fangled drift of globalisation. By adopting the ICH guidelines, the process of new drug approval can be rationalised.

REFERENCES

1. Madle, S., Korte, A., Ball, R. 1987. Experience with mutagenicity testing of new drugs: View point of a regulatory agency. *Mutat Res* 182:187–192.
2. OECD (Organisation of Economic Cooperation and Development). 1996. *Guidelines for Testing of Chemicals, Updated Guideline 473, Genetic Toxicology: In Vitro Mammalian Chromosome Aberration Test.* Organisation of Economic Cooperation and Development, Paris.
3. OECD (Organisation of Economic Cooperation and Development) 1998. *Ninth Addendum to the OECD Guidelines for the Testing of Chemicals.* Organisation of Economic Cooperation and Development, Paris.
4. Wassom, J. 1992. Origins of genetic toxicology and the environmental Mutagen Society. *Environ Mol Mutagenesis* 14:1–6.
5. Tennant, R., Margolin, B., Shelby, M., et al. 1987. Prediction of chemical carcinogenicity in rodents from in vivo genotoxicity assays. *Science* 236:933–941.
6. Purves, D., Harvey, C., Tweats, D., et al. 1995. Genotoxicity testing: Current practices and strategies used by the pharmaceutical industry. *Mutagenesis* 10:297–312.
7. Kennedy, T. 1997. Managing the discovery/development interface. *Drug Disc Toda* 2:436–444.
8. Nath, J., Krishna, G. 1998. Safety screening of drugs in cancer therapy. *Acta Haematol* 99:138–147.
9. Eaton, D., Klaassen, C. 1996. Principles of toxicology. In: Klaassen, CD (ed) *Casarett and Doull's Toxicology. The Basic Science of Poisons*, 5th edition. McGraw-Hill, New York, NY. 13–33.
10. Kushner, G.J., Silverman, R.S., Steinborn, S.B., et al. Requirement and guidelines on clinical trials for import and manufacture of new drugs. 1999. In: Malik, V (ed) *Drugs and Cosmetic Act, 1940*, 12. Eastern Book Company, Lucknow. 394–404.
11. Hutt, P., Merril, R. 1991. *Food and Drug Law: Cases and Materials.* Foundation Press, Mineola, NY.
12. EC (European Community) 1987. Notes for the Guidance for the Testing of Medicinal Products for Their Mutagenic Potential. *Official Journal of the European Community.* 73.
13. Glocklin, V. 1984. Current considerations about the role of mutagenicity studies. Drug Safety Evaluation. In: *Critical Evaluation of Mutagenicity Tests.* MMV, Medzin Verlag Munchen. 527–531. https://books.google.co.in/books?id=AfBY56itO98C&pg=PA470&dq=Glocklin,+V.+1984.+Current+considerations+about+the+role+of+mutagenicity+studies.+Drug+Safety+Evaluation.+In:Critical+Evaluation+of+Mutagenicity+Tests+.+527%E2%80%93531.&hl=en&sa=X&ved=2ahUKEwigzezkwLrqAhUt73MBHZxvDGcQuwUwAHoECAIQBg#v=onepage&q=Glocklin%2C%20V.%201984.%20Current%20considerations%20about%20the%20role%20of%20mutagenicity%20studies.%20Drug%20Safety%20Evaluation.%20In%3ACritical%20Evaluation%20of%20Mutagenicity%20Tests%20.%20527%E2%80%93531.&f=false.
14. Sofuni T. 1993. Japaneese guidelines for mutagenicity testing. *Environ Molec Mutagen* 21:2–7.
15. HPB (Health Protection Branch). 1993. Health Protection Branch Mutagenicity Guidelines. Health Protection Branch Genotoxicity Committee. Environ Mol Mutagen 15–37.

16. Cartwright, A., Mathews, B. 1994. *International Pharmaceutical Product Registration*: *Aspects of Quality, Safety and Efficacy*. Ellis Horwood Limited, New York, NY.

17. Muller, L., Kasper, P., Madle, S. 1991. The quality of genotoxity testing of drugs. Experiences of a regulatory agency with new and old compounds. *Mutagenesis* 6:143–149.

18. Scott, D., Galloway, S., Marshall, R., et al. 1991. Genotoxicity under extreme culture conditions. A report from ICPEMC Task Group 9. *Mutat Res* 257:147–204.

19. Christan, M. 1997. Overview of the fourth international conference on harmonization. *Int J Toxicol* 16:659–668.

20. Muller, L., Kikuchi, Y., Probst, G., et al. 1999. ICH harmonized guidances on genotoxicity testing of pharmaceuticals: Evolution, reasoning and impact. *Mutat Res* 436:195–225.

21. Alden, C.L. 2000. Safety assessments for non genotoxic rodent carcinogens: Curves, low-dose extrapolations, and mechanisms in carcinogenesis. *Human Expt Toxicol* 19:557–560.

22. Kim, B., Margolin, B. 1999. Prediction of rodent carcinogenicity utilizing a battery of *in vitro* and *in vivo* genotoxicity tests. *Environ Molec Mutagen* 34:297–304.

23. Brusick, D. 1987. *Principles of Genetic Toxicology*, 2nd edition. Plenum Press, New York, NY.

24. Galloway, S. 1994. Report of the International Workshop on Standardisation of Genotoxicity Test Procedures. *Mutat Res* 312:195–322.

6 Genotoxicity Prediction; Computational Prediction

INTRODUCTION

Determining genotoxic potential is a requisite for the advancement of any pharmaceutical drug. The Ames test, which measures hereditary mutagenic events in bacterial DNA, is one of the most widely used methods for the determination of mutagenic activity and remains the main explicit *in-vitro* test for the prediction of rodent carcinogenicity [1].

Most Ames positive compounds are rodent carcinogens. Ames test is sensitive because mammalian carcinogenicity is a multifaceted process whereby chemical reactivity is just amongst the many factors that give rise to tumourigenesis. Therefore, data from chromosome aberration (CA) tests must be submitted first in human trials. The ability to envisage the outcome of these assays is key to ensure for the drug concern due to which genotoxicity is minimised during the later stages of the drug discovery and development process [1].

Several *in-silico* models are available for such genetic toxicity studies, most of which are based on existing *in-vitro* test data, which have been pooled together over the years. These systems comprise two main sub-groups,

- *Expert systems* – the model is developed from human-derived structure–activity relationship (SAR), or
- *Quantitative structure–activity relationship (QSAR) systems* – the model is developed through a computer algorithm without human bias.

Ames mutagenicity is one of the easiest toxicological endpoints to predict the activity, given that it is based on a primordial cellular system and that activity in this assay is principally dependent on chemical reactivity. On the other hand, the mechanisms by which CAs occur are varied and can be composite. Indeed, CAs can be either structural or numerical, which may occur through disruption of the mitotic machinery during cell division. This results in aneugenicity [1].

On the other hand, double-strand breaks and aneugenic activity can be detected in an *in-vitro* micronucleus assay. Resulting from DNA damage, mutagenic and clastogenic events are of high concern. This leads to an increase in the risk of carcinogenic potential. In contrast, aneugenicity is related to pharmacological activity, and an agreeable margin of safety for these compounds is possible by taking into account the daily allowable intake [2].

MUTAGENICITY

QSAR Approaches

QSAR models for mutagenicity prediction may be either of the following types:

- Global QSARs are advantageous in the manner that they provide one-stop access to a mutagenicity prediction and often provide some sort of quantitative level of confidence in a prediction. Global SAR wherein entire data sets of similar and dissimilar chemicals are modelled.
- Local QSAR wherein a QSAR is used to refine the prediction of a congeneric class of compounds, such as those containing an aromatic nitro group.

This confidence in the prediction is a merit of QSAR over other expert approaches in the manner that it gives the user a concrete clutch on the accuracy of the prediction. For expert systems based purely on structural alerts, it is unclear whether the lack of a structural alert is a negative prediction for mutagenicity or a query compound simply lies outside the chemical space on which the model was developed. The ability to describe the applicability domain of a QSAR model is significant as it is a suggestion of the consistency of predictions made by the system [3,4]. The ability to build a QSAR model that is predictive of the training set is not a challenge; a good model may simply be overfitted [4].

Overfitting is a statistically complex issue which is related to predicting training set with specific features to improve predictive performance, but if it is not applicable to other data sets then it will be having a reduced applicability and performance.

The challenge before QSAR methodology is to build models that can be used for hypothesis generation and mechanistically interpreted. This may be difficult if the descriptors are not readily interpreted in a physicochemical or mechanistic sense on which the model is based. This lack of interpretability is alleviated by systems which qualify the prediction result. This is done by using a similarity algorithm to identify structurally similar compounds.

This allows the user to impart additional expert reasoning for determining whether the prediction is relevant. Examples of such software include:

- Lazar [5]
- CAESAR (http://www.caesar-project.eu).

Lazar is, for the most part, useful as it provides a direct hyperlink to a computerised and mechanised search of the DSSTox and PubChem websites. Commercially available software include ACD/Tox (http://www.acdlabs.com/ home), MCASE/MC4PC [6] and SciQSAR [6]. The Leadscope Model Applier is a new *in-silico* system designed to house the QSAR models developed by the Informatics and Computational Safety Analysis Staff (ICSAS) group [6].

CHROMOSOME DAMAGE

The prediction of chromosomal damage is a larger challenge compared to mutagenicity because the mechanisms involved are more varied and complex. Evidence suggests that such mechanisms are based on pharmacological activity and, therefore, they simply may not be modelled using QSAR strategies or through structural alerts [7]. Nevertheless, a topological sub-structural molecular design (TOPS-MODE) approach has been developed for a classification model from a data set.

- Before modelling, the data set is to be divided into both training and test sets.
- The evaluation of the resulting model gave a good predictive performance.
- The contribution of bonds to clastogenic activity is evaluated using an orthoganalised classification model; the structural alerts were identified for clastogenicity. These alerts act through a DNA alkylation mechanism.

The Derek system has structural alerts for chromosomal damage *in vitro*, which are mostly based on clastogenic compounds, but several aneugenic structural alerts have also been included. Several mechanisms of chromosomal damage have been firmed from inhibition of DNA synthesis and repair to thiol reactivity and energy depletion. Despite the considerable knowledge contained, alerts have been derived from industrial chemicals, which have an impact on the prediction of the drug-like compounds tested in the *in-vitro* micronucleus [7]. Eight MDL QSAR models have been used for the prediction of genotoxicity using the ICSAS genetic toxicity database. Fifty percent have been used for the detection of various mutagenicity endpoints and 50% for both *in-vivo* and *in-vitro* assays to detect chromosomal damage. Each model has been developed on the assay data available and evaluation is done using a leave group out internal validation strategy.

In general, predictive performance has been lower for models developed on data for chromosome damage *in vitro* than those based on Ames mutagenicity. Furthermore, sensitivity levels were considerably lower for models which were based on *in-vivo* data. This reflects supplementary attenuating factors in *in-vivo* systems, such as xenobiotic metabolism and clearance. These performance figures for such complex endpoints are extremely encouraging and their usage is evident [7].

To understand the structural requirements for non-mutagenic CAs, first, the existing mutagenicity is modified and used to classify clastogens which act through a mutagenic mechanism. From the remaining clastogens, structural alerts are then identified and moderated by physicochemical properties, which is similar to the approach used by the TIMES model [8].

By first evaluating chemicals that do not undergo bioactivation, the performance is well against the training set, which will be of 80% accuracy. The combined use of a metabolic simulator for the evaluation of the clastogenic activity of compounds requires metabolic activation and is not successful. There are chances for increased rate of false-positive predictions. Compounds are potential intercalators if they test positive in either the bleomycin assay or the docking model. Consequently, more than

50% of clastogens are predicted to be intercalators [8]. The confrontation in predicting clastogenicity is enlarged when modelling aneugenicity.

Clastogenicity and aneugenicity are simply stimulated through interruption of cellular signalling pathways. The computational model development based on kinase inhibition is used for the prediction of the chromosomal damaging activity. Thus, there is a growing accord that these kinase targets and other signalling pathways contribute to the genotoxicity of drug-like compounds in *in-vitro* assays [8].

STRATEGIES FOR GENOTOXICITY PREDICTION IN DRUG DEVELOPMENT

The application of computational models for the prediction of genotoxicity within drug development has been primarily constrained to the predictions of Ames mutagenicity. Recent efforts are focused on the prediction of non-DNA reactive genotoxicity, thereby examining clastogenic activity. It is an intricate matter of where to apply these tools which varies with the stages involved in the drug development process [9].

In the early stages of the development, the prime utility for genotoxicity predictions was to assist in the assortment of lead matter also referred to as drug candidates. There is an accent on the assessment of the sensitivity of the predictive *in-silico* model with a low false-positive rate. The main driving force is to analyse as many genotoxic compounds possible without discarding a potentially superior drug candidate. Genotoxic screens are being utilised by most pharmaceutical companies, such as mini Ames assay, in the development process. All candidates are tested for genetic toxicity as part of the standard investigation new drug package requirements which ensures prior identification of the genotoxic compounds before entering into the clinical development stages [9].

In the later developmental stages, the use of genotoxicity predictions focuses on the evaluation of low-level impurities in the active ingredient, rather than looking at the parent drug itself. Therefore, it is desirable to have *in-silico* predictive models that err on the side of caution, which minimises the number of missed potentially genotoxic compounds, which is the number of false negatives. This will minimise the potential risks presented to patients and the potential for the suspension of clinical trials at the request of a regulatory authority [9].

Muller proposed five class classification schemes to decide whether an impurity possesses a high level of risk. This classification scheme is defined in Table 6.1.

Regulatory authorities have an interest in the likely application of computational models for toxicology and seek to utilise them for the identification of probable unsympathetic events that a new drug might cause in patients or for further testing of environmental contaminants. Regulatory authorities are, however, conscious of the need to place the peril to patients presented by a novel drug candidate alongside the benefits that the drug brings in curing of the medical conditions for which it is designed. Regulatory authorities predominantly use computational models for genotoxicity which help them in characterising the impending risk presented by either low-level impurity in a drug product or for the consideration of environmental hazards [10].

TABLE 6.1

Classes for the Genotoxicity Potential of Impurities in the API

Class	Genotoxicity Potential of Impurities
Class I	Impurities are well known to be both genotoxic (mutagenic) and carcinogenic
Class II	Impurities are well known to be genotoxic (mutagenic) but with unknown carcinogenic prospective
Class III	Structural alert is distinct to the structure of the API and of anonymous genotoxic that is mutagenic prospective
Class IV	Structural alert is related to the API
Class V	No structural alert or there is insufficient evidence for the absence of genetic toxicity

The recent draft of the regulatory guidance is resolute on the requirement for consideration for the potential genotoxicity risk of metabolites uniquely seen in humans. There can be noteworthy differences between human metabolism and *in-vitro* system or *in-vivo* genotoxicity studies. Therefore, there is an increasing concern that these unique metabolites have not have been adequately studied for their risks pertaining to human safety in pre-clinical studies. To date, computational models have not been applied to address this issue, but it is very likely that once the regulatory guidance is accepted, it will become another potential application of *in-silico* systems for the prediction of genotoxicity [11,12].

CONCLUSION

Both expert and QSAR methods are uniformly successful in the prediction of the mutagenic activity of compounds on which they are being modelled; however, when it comes to true external data sets, they fall short. Directed development and testing of mechanistic SARs for the compounds incorrectly predicted lead to considerable improvement in the predictive systems. Development of more mechanistic SARs can extrapolate *in-silico* models into areas where mutagenic activity has not been explored. Thus, more confidence in predictions can be achieved by the model outside the domain of applicability.

Computational strategies are applied to chromosomal damage; it is not adequate to simply relate the current paradigm of QSAR model development. Therefore, there is a need to identify toxicological knowledge gaps and conduct focused testing so that a better understanding of the mechanistic SARs can be achieved or indeed regarding the biological pathways behind the observed toxicity.

REFERENCES

1. Kirkland, D., Aardema, M., Henderson, L., et al. 2005. Evaluation of the ability of a battery of three in vitro genotoxicity tests to discriminate rodent carcinogens and non-carcinogens I. *Mutat Res* 584:1–256.
2. Bolt, H., Foth, H., Hengstler, J., et al. 2004. Carcinogenicity categorization of chemicals-new aspects to be considered in a European perspective. *Toxicol Lett* 151:29–41.

3. Dragos, H., Alexandre, V. 2009. Predicting the predictability: A unified approach to the applicability domain problem of QSAR models. *J Chem Inf Model* 49:1762–1776.
4. Weaver, S., Gleeson, M. 2008. The importance of the domain of applicability in QSAR modelling. *J Mol Graph Model* 26:1315–1326.
5. Zhang, Q., Aires de Sousa, J. 2007. Random forest prediction of mutagenicity from empirical physicochemical descriptors. *J Chem Inf Model* 47:1–8.
6. Kazius, J., McGuire, R., Bursi, R. 2005. Derivation and validation of toxicophores for mutagenicity prediction. *J Med Chem* 48:312.
7. Hansen, K., Mika, S., Schroeter, T. 2009. Benchmark data set for in silico prediction of Ames mutagenicity. *J Chem Inf Model* 49:2077–2081.
8. Benigni, R., Bossa, C. 2008. Predictivity of QSAR. *J Chem Inf Model* 48:971–980.
9. Johnson, S.R. 2008. The trouble with QSAR (or how I learned to stop worrying and embrace fallacy). *J Chem Inf Model* 48:25–26.
10. Helma, C. 2006. Lazy structure-activity relationships (lazar) for the prediction of rodent carcinogenicity and Salmonella mutagenicity. *Mol Divers* 10:147–158.
11. Matthews, E., Kruhlak, N., Cimino, M. 2006. An analysis of genetic toxicity, reproductive and developmental toxicity, and carcinogenicity data: II. Identification of genotoxicants, reprotoxicants, and carcinogens using *in silico* methods. *Regul Toxicol Pharmacol* 44:97–110.
12. Contrera, J., Kruhlak, N., Matthews, E., et al. 2007. Comparison of MC4PC and MDL-QSAR rodent carcinogenicity predictions and the enhancement of predictive performance by combining QSAR models. *Regul Toxicol Pharmacol* 49:172–182.

7 Computational Prophecy of Genotoxicity with Models

INTRODUCTION

There is a common positive perception that an absolute understanding of all toxicities of a new-fangled drug candidate is essential for its successful expansion and marketing. Fortunately, the genetic toxicity can be premeditated unswervingly by well-known and universally conventional assays, such as the Ames test which is employed for bacterial mutagenicity; chromosome aberration (CA) assays in human lymphocytes or other mammalian cells in culture; *in-vivo* cytogenetics studies; and a host of "second tier" assays which, though not at all times are uniformly concordant, but can be applied in a weight of evidence context [1,2].

Inevitably, regulatory agency directive studies [1,2] have characteristically been demeanour in the development process after the establishment of pre-clinical efficacy at the same time casing as for the general toxicology studies. However, every pharmaceutical company has an account of how safe, sound and effectual molecules have been forced out of advance development because of unforeseen genotoxicity during regulatory studies.

There is also a call for characterisation of the genotoxic capability of metabolites, degradants, and impurities. Today, virtually all big pharmaceutical companies have developed early-on gene-tox screening programs, which usually employ a scaled-down 'mini'-Ames and an *in-vitro* appraisal of chromosomal damage in cultured mammalian cells.

Thus, genotoxicity is investigated away early on and structure–activity relationship (SAR) techniques can typically steer consequent chemical syntheses to shun genotoxicity. Most big pharmaceutical companies utilise computational programs to assist in the prediction of genotoxicity, and an amalgamation of *in-vitro* screening and *in-silico* investigation is extensively used.

Genotoxicity should be easier to envisage than other types of toxicity studies because it typically arises from direct chemical/DNA interface reliant to a larger degree on electrophilicity. Explicit organ toxicities can arise by numerous pharmacological or chemical mechanisms not inevitably related to or apparent from chemical structure analysis. The development of microarray technologies has made it possible to institute definite organ toxicity gene expression signatures, which may help predict organ toxicities necessitating longer-term pre-clinical animal studies.

Alas, even genotoxicity has demonstrated to be significantly obstinate to predict based on two-dimensional structure analysis despite the subsistence of computational programs whose "astuteness" is based on exceptionally great number of compounds and genotoxicity data.

This chapter, therefore, focuses on the prime computational programs which are used as models and their features in predicting genotoxicity. The inherent benefits and limitations of these programs and vistas for improvement have been discussed in detail for in-depth understanding.

PRINCIPAL THESPIANS

The following are the principal players with their general description:

- DEREK; Deductive Estimation of Risk from Existing Knowledge
- MCASE; Multiple Computer Automated Structure Evaluation
- TOPKAT; Toxicity Prediction by Komputer Assisted Technology
- QSAR models
- DNA Docking Model
- Consensus Modelling

DEREK SYSTEM DEDUCTIVE ESTIMATION OF RISK FROM EXISTING KNOWLEDGE

- It has been created by Lhasa Ltd. DEREK is an information- and statute-based expert system that formulates semi-quantitative inferences as to whether a DNA reactive (which is subdivided as general genotoxic, mutagenic or chromosome damaging) moiety is present on the contributing chemical structure [3,4].
- A knowledgeable user is competent to resolve if a flagged alert is, in the proper chemical context, a genotoxic relative to the compound(s) upon which the DEREK rule is based. The learning set for DEREK utilises both bacterial mutagenicity and erstwhile available genotoxicity data.
- Query productivity defines the structural alert recognised, the category of genotoxicity (bacterial mutagenicity, *in vitro* cytogenetics, etc.) associated with the alert, unambiguous examples of genotoxic compounds partaking the alerting moiety, comprehensive mechanistic comments pertinent to the alert and literature references. DEREK can be customised by the user [3,4].

MCASE (MULTIPLE COMPUTER AUTOMATED STRUCTURE EVALUATION)

- MCASE separates each input molecule into 2–10 atom wreckage, and statistical evaluation is done based on the strength of alliance of fragments known as biophores, as well as similar fragments from its database, with a coupled mutagenicity score, which is a value based on the observed mutagenic potency. It engenders a quantitative prediction of mutagenicity which is then refined additionally by considering physicochemical properties, as well as the continuation of potential "deactivating fragments", also referred to as biophobes.

- The inventive and creative MCASE model is based on bacterial mutagenicity statistics drawn from compounds from the National Toxicology Program (NTP) of the U.S. Environmental Protection Agency (USEPA), Genetox programs and, thus, pharmaceuticals were later negative in the Ames test.

- Recent version has the basis of 3000 compounds and is inclusive of Drosophila mutation data. Another version built up by the FDA in collaboration with MCASE where 16 separate modules permit the prediction of mutagenicity in individual *Salmonella* strains in the presence and absence of either rat or hamster S9 activating systems. MCASE can be customised by the user.

TOPKAT (TOXICITY PREDICTION BY KOMPUTER ASSISTED TECHNOLOGY)

- TOPKAT exercises "electro-topological" descriptors rather than chemical structures to predict mutagenic reactivity with DNA and, as such, is a conservatory of traditional quantitative structure–activity relationship (QSAR) analysis. The astuteness of TOPKAT was derived solely from bacterial mutagenicity data [3,4].

- TOPKAT was originally designed by Health Systems Inc. and is marketed by Accelrys, San Diego, CA, USA. The Ames forecast module comprises 1866 compounds which are divided into individual models based on chemical class correlation.

- Unlike DEREK, TOPKAT endows with a measure of the resemblance flanked by a test molecule and the chemical space enclosed by the program is exclusive from auxiliary analysis of any molecule deemed to have scarce coverage. TOPKAT cannot be modified and tailored by the user [3,4].

QSAR MODELS

In addition to the above programs, several QSAR models have been thought out and projected [5–12]. QSAR models use algorithms that are based on diverse chemical descriptors, such as chemical sub-structure, logP, electronics, geometrical attributes and surface area, to acquiesce an extrapolative assessment. Most QSAR genotoxicity model forecasts are based on bacterial mutagenicity data [9], which foretells and is based exclusively on CA data. Extraordinarily, this CA QSAR model is obligatory for three topological descriptors for prediction.

At present, only one QSAR model, CSGenotox (www.ChemSilico.com), has been evaluated in side by side trials with other computational programs tested against a common tester set of molecules to establish comparative performance [7]. The study compared the predictive capability of three QSAR models to that of MCASE and

DEREK for non-drugs and drugs. Of the descriptors found to be predictive and extrapolating, 40% were correlated to well-known structural genotoxicity alerts.

In line with the results, the QSAR loom had improved and enhanced specificity, but worn out unsubstantiated and unverified out of the box calls for MCASE and DEREK for evaluation and similarity, which biases the results. On the other hand, QSAR advances recommend a singular way of categorising novel chemical entities. Additionally, assessment and appraisal may prove QSAR modelling to be of better quality or at least harmonising to the non-QSAR models.

DNA DOCKING MODEL

Another approach under development is a three-dimensional DNA docking model for the recognition of molecules capable of non-covalent DNA interaction, that is, groove binding, either DNA intercalation or both [13–15]. This model determines the knack of a molecule to fill up the room between two contiguous DNA base pairs and estimates the vigour and potency of that fit based on hydrogen bonding and van der Waals forces. Preliminary studies have established and verified the utility of this model in envisaging the genetic toxicity of non-covalent DNA binding molecules [16,17].

Consequent work has revealed that many non-structurally alerting clastogenic drugs, not envisaged by customary computational programs, may act via DNA intercalation.

CONSENSUS MODELING

Based on the genetic toxicity prophecy on the combined yields of multiple models – consensus modelling – by and large improve the precision of prediction, but at the cost of reduced chemical coverage. It has been very well established and reported that [18], CASETOX, TOPKAT and DEREK have a high quality of 78% to 82% when used as solitary programs.

Combining one or both additional programs, the specificity amplified to 92% to 100% (with no substantive alteration in sensitivity), but the number of molecules that could be assessed by multiple programs (based on the criterion that both programs had to have made the same call on a particular molecule) dropped by 60% to 80%. Consensus modelling from multiple QSAR modules has also been reported to provide additional predictive power when applied to generalised polycyclic aromatics [19] or thiophenes [20].

DRAWING A PARALLEL PROPHECY

Precise and accurate prediction of Ames mutagenicity is indispensable for drug development, with the exception of drugs being developed for life-threatening diseases or conditions. A positive Ames test is more often than not the death knell for a molecule. However, there is an essential equivalent for early genetic toxicity prediction of *in-vitro* and *in-vivo* cytogenetics, the additional and supplementary two components of the regulatory testing battery. There is no rationale to consider that an optimistic computational prediction based on Ames' SAR would not be relevant and appropriate to other genetic toxicology endpoints because all true genetic toxicity is based on the identical competence for covalent DNA addition, maybe in bacteria, mammalian cells, in culture or *in vivo*. Indicating this is the substantial and significant overlap between Ames positivity and the positivity in other genetic toxicity assays among marketed pharmaceuticals [21].

Taking into consideration the foremost computational programs, structure/ genetic toxicity data other than Ames results have been built only into DEREK, though novel modifications are available through MCASE versions. These versions are being developed as unambiguous and explicit models for forecasting mouse lymphoma and CA assays. The QSAR model [9] based only on CA data. Therefore, it is very interesting to perceive and distinguish if these broadly based models will be more extrapolative and predictive for either comprehensive or explicit genetic toxicity endpoints.

SENSITIVITY

Sensitivity is the appraisal of a program's capability to recognise true positives in an approved manner. Lack of sensitivity is a problem that arises only if one concentrates too profoundly on the computational analysis to press forward for a compound to the candidate stage or in the creation of hand safety pronouncements. However, in the non-existence of a computational prediction of genetic toxicity, all probability weigh the genetic toxicity of any fragment of significance predominantly if the molecule is infatuated with a structural module of apprehension.

Therefore, lack of sensitivity may be more of an academic inquisitiveness than a drug development stumbling block. On the other hand, a more inclusive and absolute consideration of the nature of false negatives is obligatory to the progress of computational models. As pointed out earlier, the unswerving comparison of models is quite intricate because, for the most part, models have not been analysed side by side with the same analysis sets.

It is, therefore, very comprehensible and apparent that all the three programs are approximately equivalent (deprived) regarding sensitivity for detecting Ames positives. This has been reported in a range of proprietary pharmaceuticals in which all three programs were confronted with a tester set of proprietary pharmaceuticals [3,18].

The sensitivities in non-bacterial genetic toxicity assays are of inferior quality than for Ames test for all the three programs. An awe-inspiring preponderance of accurate and precise calls for all assay types by all three programs was completed on drugs with understandable and apparent structural alerts as at the beginning defined by Ashby [23,24].

This is to be predictable as each computational system has these structural patterns controlled within their respective training sets. Alternatively, the uppermost pragmatic sensitivity value for non-Ashby alerting PDR drugs indicated that these programs were not skilled regarding non-alerting drugs. From within the PDR test set, non-alerting genotoxic drugs have been acknowledged [22]. They cooperatively add up for all positive Ames, *in-vitro* cytogenetics, mouse lymphoma and *in-vivo* cytogenetics results.

Some of the molecules might have failed to spot, which may be due to the known obligation for metabolic activation. Some molecules have failed to spot because they were conventional intercalating agents, information on which is underrepresented in these models. Some drugs have also failed to spot for which the reason is unknown, however, latest studies have suggested that numerous structurally diverse molecules are competent of functional DNA intercalation even though they do not have conventional fused ring or planar intercalating structures [16,17]. Because only a diminutive number of classical and non-classical intercalating structures are incorporated in the databases of these programs, it would give the impression that additional characterisation of non-covalent DNA interaction and addition of this in sequence into computational models extent amplify and augment the sensitivity to a large.

However, non-covalent DNA binding possibly will even additionally be imperative and significant in the circumstance of positive CA effects because intercalating agents are well-known topoisomerase inhibitors and, as such, engender largely irretrievable DNA double-strand breaks [25].

To some extent, the sensitivity of these computational models is enhanced if one tests chemicals relatively than drugs [26]. In all probability, as drugs are underrepresented in these models, the chemical space or degree of chemical diversity corresponded to or signified by the molecules worn out to construct each program is augmented with non-pharmaceutical-type structures. The involvedness of using these programs based on a comprehensive non-congeneric set of molecules has also been discussed in the QSAR context [11]. The equivalent is true of the databases that are created from a set which is structurally related to the proprietary molecules, which is then confronted with supplementary structures within that equivalent chemical space. Thus, these programs are innately more accurate, precise and exact when testing structurally confined rather than global tester sets.

SPECIFICITY

The specificity of the three primary programs with deference to their overall recital and performance characteristics can be confronted with PDR drugs. The MCASE and DEREK values are drawn from employing the connoisseur assessment of the program outputs, and not out of the box calls. Without client input, recital characteristics are considerably underprivileged. Overall, concordance is a whole percentage of accurate positive and negative calls, is often hoodwinking and is more often than not is biased by the superior quantity of true negatives in most tester sets. High concordance contradicts and disapproves the compassion of all of these programs. Out of the way from the underprivileged sensitivity, it is clear that specificities could, in addition, make use of some improvements.

Specificity is, thus, defined as the ratio of true negatives to the sum of true negatives and false positives. Thus, for instance, if there are 50 true negative forecasts out of which there are 25 false-positive forecasts, the specificity would be 67%, which is unacceptably very low.

The cohort of false positives can be a consequence in the discarding of useful drugs. This would in all probability not be a concern in a combinatorial chemistry program where an assured percentage of molecules must be sorted out at any rate to edge out the number of molecules that are taken to the subsequent discovery step. However, if the computational investigation and scrutiny are done afterwards when, for example, 20 compounds are being rank-ordered, throwing away false positives can be more challenging and difficult. In addition, a false positive genetic toxicity prophecy, even if is tested unswervingly by bioassay and initiated to be clean, might require some sort of supplementary testing or a terse discussion with the regulatory agencies. This would be predominantly true for molecules clutching structural alerts.

WHAT CAUSES THIS HIGH NUMBER OF FALSE POSITIVE PREDICTIONS?

Available declarations and assertion regarding to the tremendously reduced specificities of computational programs are generally referred to as out of the box calls, and on odd occasions take expert input into the consideration. As stated above, false positives could be markedly trimmed down by including an expert opinion as to whether the recognised biophore is in a chemical context unswerving with it being DNA reactive. This fortitude can, by and large, be prepared by a chemist or toxicologist. However, there are other sources of false-positive calls. Inventors and developers of MCASE have reported that extrapolative value can be reliant on the following:

- The integer of molecules,
- The proportion of positive to negative compounds, [27,28] and
- The nature of the molecules, for example, drugs versus pesticides, in the learning data set [29].

In addition, the poor specificity could also be due in part to the inaccurate handing over of scrap as being conscientious for the mutagenicity of a molecule that has nix true biophore. For example, a mutagenic nucleoside would not hold a reliable structural alert, its mutagenicity as a replacement for being due to reticence of DNA polymerisation. Correlative looms such as MCASE would, on the other hand, seek out to classify a portion(s) to lay at the access of mutagenicity and such obligations would be inevitably erroneous.

Consequent test molecules hauling this same scrap may be erroneously recognised as positives. Thus, there is the removal of non-alerting positive gene-tox structures from the MCASE learning set and further confronted with the PDR database. Although some upgradation and enhancement of specificity (that means lesser false positives) did come about, it is comprehensible that such erroneous biophores account for only a diminutive portion of the false positives seen with MCASE.

GENOTOXIC POTENCY

The PDR endows with the kind prospect in order to estimate and appraise the genetic toxicity results obtained from complemented study blueprint and then focused on the identical interpretative criterion. The database is also innately undermined, however, by the deficiency in the information regarding the potency of the response and the dose at which the genetic toxicity response was scrutinised.

This is an attribute regrettably collected by many widely accessible and obtainable toxicology database collections. In recent times, at least two self-determining proposals have tried to concentrate on this limitation in appropriate and apposite SAR data. The primary of these is VITIC, which is a database concept primarily produced from a Health and Environmental Science Institute (HESI) enterprise in collaboration for development with Lhasa where publicly accessible toxicology statistics are offered in a database set up with rummage around chemistry and its links to the inventive and creative biological assay data.

The second is the DSSTox, a record notion title holder by Ann Richard of the USEPA, in which statistics and records are shared as SDF files in a regular ASCII format [30].

It is expected that these potency records will support in the development of risk estimation paradigms. Most of these genotoxic responses are expected to be very feeble and frail and to take place at concentrations far ahead of those targeted in the workshop. These optimistic and affirmative responses, though approximately and indeed genuine, in all probability carry minimum to negligible menace to the patient population and an enhanced way of appraising the tangible and authentic peril coupled with these drug exposures would be highly enviable and sought after [21]. Thus, it can be understood that carcinogenicity arises via additional genotoxic mechanisms. Thus, in continuation, it can also be suggested that feeble and weak genetic toxicity does not inevitably transform into carcinogenicity. This is mostly in all probability because of exposure deliberations, rather than an erroneous genetic toxicity result, but this has not been meticulously and scrupulously evaluated. Such study reports a lucid and comprehensible linear relationship as observed between *in-vivo* genetic toxicity potency and dose-defined carcinogenic [31].

Thus, measuring feeble versus sturdy *in-vitro* genotoxins with deference to drug plasma levels and carcinogenicity outcome in the bioassay might be further helpful. Such studies would be ready to lend a hand in deciding if cogent and coherent risk appraisal could be based on potency that means DNA reactivity and clinical plasma drug concentration.

CONCLUSION

The currently accessible *in-silico* loom regarding genetic toxicity prediction appends a value to the drug development process. However, it is apparent that the existing tools do not provide the desired degrees of both sensitivity and specificity. Although noticeable and discernible progress in concordance can be made via expert interpretation of the out of the box calls, each computational model still necessitates wide-ranging modification before the eventual objective of reinstating biological assays can be accomplished. It is, thus, anticipated that one area in which noteworthy step-ups and perfection can be made is in the reporting of non-bacterial genotoxic responses. In particular, enclosure of specific genetic toxicity data applicable and relatable to non-covalent DNA interactions can show the way to a discernible amplification in the prediction assessment and significance.

REFERENCES

1. ICH S2A: Guidance on specific aspects of regulatory genotoxicity tests for pharmaceuticals, CPMP/ICH/141/95.
2. ICH S2B: Genotoxicity: a standard battery for genotoxicity testing for pharmaceuticals, CPMP/ICH/174/95.
3. Green, N. 2002. Computer systems for the prediction of toxicity: An update. *Adv Drug Deliv Res* 54:417–431.
4. Cariello, N. 2002. Comparison of computer programs DEREK and TOPKAT to predict bacterial mutagenicity. *Mutagenesis* 17:321–329.
5. Votano, J. 2004. Three new consensus QSAR models for the prediction of Ames genotoxicity. *Mutagenesis* 19:365–377.
6. Basak, S. 2001. Prediction of mutagenicity of aromatic and heteroaromatic amines from structure: A hierarchical approach. *J Chem Inf Comput Sci* 41:671–678.
7. Votano, J. 2005. Recent use of topological indices in the development of *in silico* ADMET models. *Curr Opin Drug Discov Dev* 8:32–37.
8. Mattioni, B. 2003. Predicting the genotoxicity of secondary and aromatic amines using data subsetting to generate a model ensemble. *J Chem Inf Comput Sci* 43:949–963.
9. Serra, J. 2003. Development of binary classification of structural chromosome aberrations for a diverse set of organic compounds from molecular structure. *Chem Res Toxicol* 16:153–163.
10. Benigni, R., Giuliani, A. 1996. Quantitative structure activity relationship (QSAR) studies of mutagens and carcinogens. *Med Res Rev* 16:267–284.
11. Benigni, R., Richard, A. 1996. QSARs of mutagens and carcinogens: Two case studies illustrating problems in the construction of models for non congeneric chemicals. *Mutat Res* 371:29–46.
12. Livingstone, D. 2002. Modeling mutagenicity using properties calculated by computational chemistry. *SAR QSAR Environ Res* 13:2–33.
13. Hendry, L. 1994. Design of novel anti estrogens. *J Steroid Biochem Mol Biol* 49:269–280.
14. Hendry, L. 1998. The ligand insertion hypothesis in the genomic action of steroid hormones. *J Steroid Biochem Mol Biol* 65:75–89.
15. Hendry, L. 1999. Multidimensional screening and design of pharmaceuticals by using endocrine pharmacophores. *Steroids* 64:570–575.
16. Snyder, R. 2004. Evaluation of DNA intercalation potential of pharmaceuticals and other chemicals by cell based and three dimensional computational approaches. *Environ Mol Mutagen* 44:163–173.
17. Snyder, R., McNulty, J., Zairov, G., et al. 2005. The influence of N dialkyl and other cationic substituents on DNA intercalation and genotoxicity. *Mutat Res* 15:88–99.
18. White, A. 2003. A multiple *in silico* program approach for the prediction of mutagenicity from chemical structure. *Mutat Res* 539:77–89.
19. He, L. 2003. Predicting the genotoxicity of polycyclic aromatic compounds from molecular structure with different classifiers. *Chem Res Toxicol* 16:1567–1580.
20. Mosier, P. 2003. Predicting the genotoxicity of thiophene derivatives from molecular structure. *Chem Res Toxicol* 16:721–732.
21. Snyder, R., Green, J. 2001. A review of the genotoxicity of marketed pharmaceuticals. *Mutat Res* 488:151–169.
22. Snyder, R. 2004. Assessment of the sensitivity of the computational programs DEREK, TOPKAT and MCASE in the prediction of the genotoxicity of pharmaceutical molecules. *Environ Mol Mutagen* 43:143–158.
23. Ashby, J. 1989. Classification according to chemical structure, mutagenicity to Salmonella, and level of carcinogenicity of a further 42 chemicals tested for carcinogenicity. U S National Toxicology Program. *Mutat Res* 223:73–103.

24. Ashby, J., Tennant, R. 1991. Definitive relationships among chemical structure, carcinogenicity and mutagenicity for 301 chemicals tested by the U.S.N.T.P. *Mutat Res* 257:229–306.
25. Ferguson, L. 1998. Inhibition of Topo II enzymes: A unique of environmental mutagens and carcinogens. *Mutat Res* 400:271–278.
26. Snyder, R. 2003. A cross-platform comparison of *in silico* models for predicting genotoxicity using marketed pharmaceuticals. *Environ Mol Mutagen* 41:207.
27. Rosenkranz, H., Cunningham, A. 2001. SAR modeling of unbalanced data sets. *SAR QSAR Environ Res* 12:267–274.
28. Rosenkranz, H.S. 2004. SAR modeling of genotoxic phenomena: The consequence on predictive performance of deviation from a unity ratio of genotoxicants/non-genotoxicants. *Mutat Res* 559:67–71.
29. Rosenkranz, H., Cunningham, A. 2001. SAR modeling of genotoxic phenomena: The effect of supplementation with physiological chemicals. *Mutat Res* 476:133–137.
30. Richard, A., Williams, C. 2002. Distributed structure-searchable toxicity (DSSTox) public database network: A proposal. *Mutat Res* 499:27–52.
31. Sanner, T., Dybing, E. 2005. Comparison of carcinogens and in vivo genotoxicity potency estimates. *Basic Clin Pharmacol Toxicol* 96:131–139.

8 Bioindicator of Genotoxicity
The Allium cepa *Test*

INTRODUCTION

Numerous species of medicinal plants have been used in popular medicine. However, they can cause public harm owing to their indiscriminate and uncontrolled use; therefore, adequate information on these plants, right from their cellular levels up to their effects on living organisms is imperative. The germplasm of medicinal species is an asset as it has tremendous potential and should be preserved and developed, making it a more reasonable substitute form of therapy that can be used by the common population, while conserving biodiversity. Studies on the categorisation and depiction of these plants on several levels, biological and/or agronomical, are indispensable and crucial, which is inclusive of the competence of their extracts depending on other living organisms [1].

Cytogenetic studies of plant species have shown possible alterations of plant chromosomes because of the mutagenic substances present either in their composition or as a result of their metabolism [1]. These methods are employed for the study of mutagens in eukaryotic nuclei. It is identified that the mutation may be caused by radiation, drugs, viruses and the intrinsic stability of nucleic acids. Therefore, cytological detection can be done by the following techniques:

- Cellular inhibition
- Disruption in metaphase
- Induction of chromosomal aberrations
- Numerical and structural
- Arraying from chromosomal splinter process to the disassociation of the mitotic spindle
- Dependent mitotic phases

Priority should be laid on the genotoxicity studies of medicinal plant extracts which will enable invested research efforts towards public health. The assessment, inspection and check of the chromosomal modifications apart from allocation as a mutagenicity test is one of the unswerving and undeviating methods, which compute and evaluate the damage in the system which is exposed to probable mutagens or carcinogens. To enable this evaluation of the damages that mutagenic agents cause, the sample needs to be in constant mitotic division, which will seek to categorise the noxious consequences and variations occurring over a cell cycle. The *Allium cepa* test has been widely used for this purpose [1].

The mitotic index/indicator and replication index/indicator are pointers of passable cell propagation and can be computed by the plant test system *Allium cepa*. Cytotoxicity tests have been *in vivo* validated using plant test systems, jointly performing animal testing *in vitro*, and thereby providing precious information for human health. The *Allium cepa* test has been used for detecting toxicity/genotoxicity as well as evaluating environmental pollution [2].

As previously mentioned, medicinal plants have been used for treating illnesses, and this test is important for alerting the population for possible genotoxic risks. This can be caused in eukaryotic plant organisms, such as onions. The focus is on the use of this test as a bioindicator of genotoxicity aiming at human health. This indirectly demonstrates that it is likely to prevent and avoid environmental contamination by the obnoxious use of substances that cause chromosomal aberrations. The applications of this test are surprising and show that certain plants are anti-mutagenic, which would allow the reversion of genotoxic processes [2].

This test is a constructive, squat outlay system, and facilitates the knowledge from plant cytogenetic techniques. The sensitivity and correlation of the *Allium cepa* (onion) test system and additional test systems are straightforward and uncomplicated for accurate appraisal of environmental problems, as well as the extrapolation and interpretation of data to other organisms such as humans [2].

DEPICTION AND MAGNITUDE OF THE *ALLIUM CEPA* TEST

The *Allium cepa* test has been used by many investigators mainly as:

- Bioindicator of environmental pollution
- Testing crude extracts of cyanobacteria
- Valuation of the genotoxic prospective of the medicinal plants

The *Allium cepa* test is imperative as it is a first-rate model *in vivo*, where the roots nurture in direct contact with the essence or agent of interest, effluent or multifaceted medicinal mix, thereby making it possible to predict the damage to the DNA of eukaryotes. This statistics can be predictive and foretelling for all animal and plant biodiversity. The investigation of chromosomal alterations can be equal to the test of mutagenicity mainly for the detection of structural alterations. It is also possible to scrutinise numerical chromosomal alterations [3–5].

The *Allium cepa* test is the unswerving technique for the measurement of the damage caused in systems which are exposed to either mutagenic substances or impending carcinogenic substances. This enables the evaluation of the effects of the damage caused by the observation of chromosomal alterations [3–5]. Therefore, the sample must remain in constant mitotic division to identify the toxic effects and alterations over a cell cycle. The *Allium cepa* test has been used extensively for this rationale. It is beneficial to use the *Allium cepa* test system as the main component is a vascular plant, which makes it as an excellent genetic model for the evaluation of environmental pollutants, detection of mutagens in diverse environments and evaluation of many genetic endpoints (point mutations to chromosomal alterations). *Allium cepa* is distinctive regarding its efficiency in detecting genetic damage. It was introduced in 1938 by Levan, in 1938, for monitoring the disturbances in the mitotic fuse due to colchicine action [3–5].

Allium cepa cells restrain an oxidase enzyme system which is competent and proficient for the metabolism of polycyclic hydrocarbonates. Though other test systems have shown to be sensitive for this detection, the results of the *Allium cepa* test are considered as an alert for other organisms (i.e. bioindicators). For an accurate evaluation of environmental risks and extrapolation of data to other groups of target organisms, there should be a fundamental focus on sensibility and correlation with and among test systems [3–5].

The use of medicinal plants for treating illnesses is an exploratory practice that is widely diffused, and because of this intense practice of medicinal use, studies using bioindicators of toxicity and mutagenicity, such as the *in-vivo* test of *Allium cepa* are indispensable for contributing to their safe and efficient use [3–5].

The plant test system of *Allium cepa* is as an ultimate bioindicator for assessing genetic toxicity, serving with the cram that averts damage to human vigour. *Allium cepa* test can be used to test infusions at concentrations of different species, thereby making it possible to verify the anti-proliferative activity of the species. The result is indicative that species possess the competence to inhibit cell division and genotoxic activity [3–5].

METHODOLOGY OF THE *ALLIUM CEPA* TEST

The *Allium cepa* test consists of acquiring onion bulbs nurtured without application of any herbicides or fungicides. The bulbs are crumbed at the root to uphold the coming out of new roots. Other bioassays can be executed with *Allium cepa* seeds for germination in a biochemical oxygen demand incubator with managed and controlled temperature, which are then worn out for allelopathy investigation as well as genetic toxicity evaluation and appraisal [6].

EXPERIMENTAL PROCEDURE

- The onion bulbs are laid initially in a small 50 ml plastic cup, which is filled with distilled or tap water (potable water) for approximately 3–4 days for the emergence of rootlets.
- The bulbs should be then transferred to other clean and dry containers. The plastic cups for water control treatment can be reused.
- In general, five groups of bulbs of *Allium cepa* are used for treatment, where one is a negative control in water and others are positive control in methylsulfonylmethane (MMS) or glyphosate. Different concentrations of glyphosate have demonstrated chromosomal alterations undeviating to the rootlets. The residual effect on the environment is not proven.
- After the emergence of roots, the bulbs of the two control groups are sustained in water as negative control and in the relevant positive control. The remaining are transferred to the chosen treatments, which can be either solution of essential oil, leaf extracts by infusion, root or stem extracts by decoction, or samples of industrial and/or hospital effluents. These are to be maintained in the dark for exactly 24 h.

Note: Ethanol PA is used as dilution medium of the oil in which the rootlets of *Allium cepa* are immersed.

- After the rootlets are maintained for 24 h in the individual treatments, they are collected and immediately fixed in a mixture of ethanol: acetic acid in a ratio of 3:1 for 24 h.
- The rootlets are then removed from the fixing solution and transferred to ethanol (70%) and are kept under refrigeration (4°C) until use. It is essential and imperative that all glassware used to carry on the rootlets ought to be acquainted with a specific number or a sample name and/or treatment, along with the date, using small labels/tags written in pencil on both the inner and outer part of the glass, which will help avoid any mistake with the samples.
- The slides are put in order, and analysis is done by analysing 1000 cells per bulb, totality to 5000 cells per treatment or disparity of these values, such as 500 cells per bulb, totalling to 2500 cells.
- One rootlet is passable and sufficient for scrutinising the damage caused to the DNA of *Allium cepa* for observing the cells after the treatment with mutagenic agents.

Note: The researcher should have a large sample so that errors can be minimised. We believe that at least 200 cells be analysed per bulb, totalling 1000 cells per treatment, in case of optimising time, considering such samples as a pilot or initial experiment.

- Slide preparation can follow the entire technique of squashing and staining of the root tip for obtaining cells with good visualisation. The hydrolysis of the rootlets is performed in hydrochloric acid for 5–6 min and rinsed in distilled water.
- Meristematic section of the rootlets is fragmented with the elimination of the rest of the part and the root cap under the stereoscopic microscope lens by employing the histological needles, which are earlier stained with 2% of acetic orcein.
- The distance with the apical meristem, to find the location with the majority of cell division varies among species, and within a single species, it depends on the age of the root. Cell division region including the quiescent region is the combination of apical meristem with the portion of the root where the cell division occurs.
- Other stains can also be used, such as acetic orcein 1%, acetic carmine 1%–2% and Giemsa. During the squashing and staining procedure, a coverslip is used which is carefully squeezed with filter paper to remove excess stain. One edge of the coverslip is knocked with the tip of the histological needle significant number of times to spread the cells. Excessive force can tear the cells and should not be confused with a morphological alteration.
- The slides are then evaluated using LEICA microscope with 400× magnification and the cells are observed during various cell division stages. Cell counts are carried out by considering visual fields scanning of the entire slide. The mitotic index (MI) is then estimated for each treatment.

- During the cell count, they are divided into two categories:
  - Regular which does not present damages in the chromosomes and
  - Irregular which present damages in the chromosome, such as chromosomal breakages, simple or multiple anaphasic bridges, micronucleus, laggard or lost chromosomes.
- The statistical analysis should be preferably performed using chi-square test with a probability level of $p < 0.05$ using a statistical program, such as BioEstat 4.0 or BioEstat 5.0.

EXPLOIT AS AN 'ADMONITION' BIOINDICATOR IN DETECTING GENOTOXICITY OF MEDICINAL PLANTS

Different species of medicinal plants are used in popular medicine for the treatment of illnesses. However, either the presence of cytotoxic and mutagenic substances in their composition or resulting from their metabolism can cause potential damage to human health. The mutagenic effects resulting in chromosomal alterations are detected during the cell cycle with the help of cytogenetic analysis [7].

What does genotoxicity mean? It is the capacity of clastogenic agents causing lesions in the genetic material. Genotoxic agents can be defined functionally as possessing the capability to alter DNA replication and, thus, genetic transmission. The assessment of genotoxicity includes:

- DNA damage
- Mutations
- Chromosomal alterations

The observation of cells in the interphase stage and cell division is considered an indicator of adequate propagation of the cells, which can be calculated through the *Allium cepa* test system [7–10].

The studies carried out with the *Allium cepa* test to torment the genetic toxicity of complex mixtures is referred to as teas or extracts. The evaluation of the genotoxic effects of the plant extracts and the sensitivity of the system have a correlation, which needs to validate its utility as a substitute test for monitoring the probable genotoxicity of environmental chemicals and pesticides [11].

Meristematic onion cells and rat cells are employed and utilised as test systems for the substantiation and subsequent authentication of the effects of the genetic toxicity of extracts (also called as infusions) of medicinal plants, which have proven that it is not worth mentioning the disparity in the fall off of the MI. There is only a decrease in the MI of the meristematic cells in the onion. These studies indicated that the use of medicinal plants could be continued as long as they are used in the recommended dosage [11].

Extracts are most commonly prepared by;

- Infusion and
- Decoction which depends on the part of the plant employed.

In *infusion*, extraction is carried out by subjecting the plant material in boiling water in a covered container for a definite time period. Infusions are applied to plant parts of the supple structure, which is beaten, cut or pulverised roughly as per their nature to enable easy penetration and then extracted with water [11].

Decoction is maintaining the plant material in contact with a boiling solvent, which is usually water, for a definite time period. It is not used widely as many active agents are distorted by extended heating, and it is, therefore, expected to be employed with rigid/woody plants [11].

Across the globe, many species of medicinal plants have been used to treat ill-nesses. Many of the species have not been thoroughly studied regarding the pres-ence of toxic/mutagenic substances either in their composition or arising from their own metabolism, which in turn damages the health. The presence of these mutagenic substances causing chromosomal alterations can be detected during the cell cycle. The *Allium cepa* test is recurrently used for evaluating the impending genotoxicity of medicinal plant extracts by analysis of meristematic cells from root-tips which are treated with medicinal infusions (teas). This knowledge of the potential genotoxicity serves as an indicator of safety for the entire population using these medicinal teas for treatment [11,12].

CATEGORY OF CONSEQUENCES AND ELUCIDATION THROUGH THE SCRUTINY OF PLANT CYTOGENETICS

The results obtained by analysing the rootlets when subjected to different treatments of interest to researchers are performed by cytogenetic analysis during cell division. The *Allium cepa* cell cycle can be taken into consideration after 24 h, and it is divided into interphase and cell division, including the prophase, metaphase, anaphase and telophase. For interpreting the results in the case of the *Allium cepa* test, the subdivi-sion is necessary [12]. As cells which are in the interphase stage are not considered as cells in division, the MI can be calculated as follows:

Mitotic Index = Total number of cells observed × {cells in interphase × Number of cells in division} / Number of cells in interphase

The slide must be put in order from the meristematic region by removal of the root cap using the squashing method immediately after stained with either 2% acetic orcein or another stain, which has demonstrated the same affinity for DNA packaged as chromosomes. The observation of the cells can thus interpret through the micro-scope based on the regular cell division of *Allium cepa* [12].

Depending on the tested substances, such as herbicides which are used in the agricultural practice or leftover drugs being discarded after use, it is possible to observe whether these substances are mutagenic or even anti-mutagenic [12]. If they are mutagenic, it is possible to immediately visualise through the structural damages where there are:

- Chromosomes with breaks
- Simple anaphasic bridge
- Multiple anaphasic bridges

- Adhesions
- Laggard chromosomes
- Disorganisation of the metaphase
- Binucleated cells

If they are anti-mutagenic, it is obligatory to verify the reversion of the mutations occurred.

The exploit of anti-mutagens and anti-carcinogens is a generally competent procedure to prevent human cancer and genetic illnesses. There are a lot of approaches in which the action of mutagens can be either decreased or thwarted. The anti-mutagenic potential of curcumin has been employed on chromosome aberrations in *Allium cepa*. Turmeric has long been used as a spice and food colouring agent. The anti-mutagenic potential of curcumin was evaluated using the root meristem cells of *Allium cepa*. The curcumin insignificantly induces chromosomal aberrations. It has an anti-mutagenic potential as well against sodium azide which is known to induce chromosomal aberrations. Thus, the mechanism of action remains unknown, though curcumin has revealed an anti-mutagenic potential, as demonstrated by the *Allium cepa* test [12].

CONCLUSION

In conclusion, the *Allium cepa* test is an exceptional bioindicator of chromosomal alterations as well as genotoxicity. Currently, due to environmental pollution concerns, the *Allium cepa* test has become imperative for the prevention and prediction of the environmental impact that will be caused by the exercise and removal of substances, including drugs and herbicides.

Although the test is merely a primary evaluation of genotoxicity, significant scientific discoveries and new adaptations of the test might reveal copious possibilities of its use, which will help in avoiding the use of animals for testing. More augmentation and investigation, as the sophistication of the method progresses, will lead us to get the most use for the benefit of the planet.

REFERENCES

1. Abdou, R., Megalla, S., Moharram, A., et al. 1989. Cytological effects of fungal metabolites produced by fungi isolated from Egyptian poultry feedstuffs. *J Basic Microbiol* 29:131–139.
2. Al Sabti, K., Kurelec, B. 1985. Induction of chromosomal aberrations in the mussel *Mytilus galloprovincialis* watch. *Bull Environ Contam Toxicol* 35:660–665.
3. Antoniou, M., Cruz, A., Dionysiou, D. 2005. Cyanotoxins: New generation of water contaminants. *J Environ Eng* 131:1239–1243.
4. Ayres, M. Ayres, M. 2003. *BioEstat 3.0: Aplicações estatísticas nas áreas das ciências biológicas e médicas*. Sociedade Civil Mamirauá, Belém.
5. Bagatini, M., Fachinetto, J., Silva, A., et al. 2009. Cytotoxic effects of infusions (tea) of *Solidago microglossa* DC. (Asteraceae) on the cell cycle of *Allium cepa*. *Braz J Pharmacog* 19:632–636.
6. Bagatini, M., Silva, A., Tedesco, S. 2007. Uso do sistema teste de Allium cepa como bioindicador de genotoxicidade de infusões de plantas medicinais. *Braz J Pharmacog* 17:444–447.

7. Bagatini, M., Vasconcelos, T., Laughinghouse IV, H., et al. 2009. Biomonitoring hospital effluents by *Allium cepa* L. test. *Bull Environ Toxicol Contam* 82:590–592.
8. Bolognesi, C., Landini, E., Roggieri, P., et al. 1999. Genotoxicity biomarkers in the assessment of heavy metal effects in mussels: Experimental studies. *Environ Mol Mutagen* 33:287–292.
9. Cabrera, G., Rodriguez, D. 1999. Genotoxicity of soil from farmland irrigated with wastewater using three plant bioassays. *Mutat Res* 426:211–214.
10. Camparoto, M., Teixeira, R., Mantovani, M., et al. 2002. Effects of *Maytenus ilicifolia* Mart. and *Bauhinia candicans* Benth infusions on onion root tip and rat bone marrow cells. *Gen Mol Bio* 25:85–89.
11. Chauhan, L., Saxena, P., Gupta, S. 1999. Cytogenetic effects of cypermethrin and fenvalerate on the root meristem cells of *Allium cepa*. *Environ Exp Bot* 42:181–189.
12. Çelik, T., Aslantürk, O. 2006. Anti mitotic and anti genotoxic effects of *Plantago lanceolata* aqueous extract on *Allium cepa* root tip meristem cells. *Biologia* 61:693–697.

9 Genotoxicity Appraisal of Nano-Sized Materials and Particles

INTRODUCTION

Nanomaterials (NMs) and nanoparticles (NPs) have diverse applications in society as well as manufacturing, which is expected to progressively augment in the future. Therefore, risk appraisal of these materials is needed. Genotoxic and mutagenic effects need to be vigilantly evaluated in relation to diseases such as cancer and inherited genetic damage. Genotoxicity caused by NPs and the underlying mechanisms have been discussed in detail along with the applicability of the different methods which are used for genotoxicity testing of NPs [1].

In particular, the use of Comet assay, micronucleus assay, chromosome aberration test, bacterial and mammalian mutagenicity tests and cell transformation assays have been described. A brief discussion is also mentioned about the possible interference of the NPs and the assays [1], some of which are underlined below:

- For the Comet assay, when high concentrations of reactive NPs are tested *in vitro*, there is a risk for DNA damage caused by additional damage.
- Micronucleus assay treatment with Cytochalasin B (to score micronucleus in once-divided cells) can affect NP uptake; therefore, delayed co-treatment is recommended.

One imperative question for all NP studies is dosimetry consideration and the fact that the real cell dose is seldom measured. Indeed, bacterial cells have a restricted ability to swallow NPs, and mammalian cells are therefore suggested for mutagenicity testing. The *in-vivo* genotoxicity studies have been compiled, including silicon dioxide (SiO_2), titanium dioxide (TiO_2), gold (Au), silver (Ag) and carbon nanotubes (CNTs) [1].

- For all materials, positive and negative studies have been reported. Remarkably, after administration via the lung, no effects on blood or bone marrow cells were observed.
- In contrast, convincing local effects in lung cells were observed for CNTs, but not for the other NPs. For TiO_2, several studies showed genotoxicity following oral exposure.
- Both Au and Ag NPs were also genotoxic subsequent on injections, and convincingly positive genotoxicity findings in a range of *in-vitro* studies were reported.

• From the *in-vivo* studies it is perceptible that the route of administration is imperative when studying the genotoxicity of NMs and a centre of attention on a target tissue is decisive.

NANOPARTICLE-INDUCED GENOTOXICITY AND UNDERLYING MECHANISMS

The detailed mechanism of NP-induced genotoxicity is not completely understood and it is unclear whether there are any nano-specific effects on DNA. The nano-specific effect refers to a mechanism of toxic action that is explicit to particles with dimensions within the size range of 1–100 nm as opposed to being associated with particles of different sizes but with the same chemical composition. In general, particle-induced genotoxicity can be categorised as either primary genotoxicity or secondary genotoxicity. Primary genotoxicity can be attributed as genotoxicity from the NPs themselves, whereas secondary genotoxicity can be attributed to the stimulation of genotoxicity via reactive oxygen species (ROS) engendered during particle-elicited inflammation [2].

PRIMARY GENOTOXICITY – DIRECT AND INDIRECT MECHANISMS

If NPs enter the karyon, either by penetration via nuclear pores or during mitosis, they might directly interact with DNA. Direct DNA interaction could represent a more nano-specific mechanism because small NPs may reach the nucleus via transportation through the nuclear pore complexes (NPC).

NPCs square measure the sole channels through which tiny polar molecules, macromolecules and NPs can travel through the nuclear envelope and it consists of a tube with a diameter of approximately 30 nm. Particles larger than 30 nm are solely transported through the pores once labelled, for example, with nuclear localisation sequence (NLS) [3].

However, larger NPs of, for example, silver (60 nm), SiO_2 (40–70 nm) and CuO (50–100 nm) have been observed in the nucleus, suggesting that larger NPs could get access to the DNA in dividing cells once the nuclear membrane disassembles. If NPs interact or bind with DNA molecules, this could influence DNA replication and transcription of DNA into RNA. To study this, Li and co-workers proposed that DNA-binding assays can be useful and showed that NPs (size range 3–46 nm) with a high affinity for DNA powerfully inhibited DNA replication (tested acellularly), whereas NPs with low affinity had no or nominal impact. Such experimental acellular studies do not consider important factors such as the ability of the NPs to enter the nucleus and the fact that DNA is highly packed in mammalian cells [3].

The probability for nuclear localisation and DNA interaction depends on the NP size similarly as its charge. For example, Nabiev et al. demonstrated that green (2.1 nm) quantum dots but not red ones (3.4 nm) entered the nucleus of THP-1 cells via NPC. One novel manner of learning direct DNA interaction leading to stalled replication forks is to use communicator cells sensitive to such effects. This approach was used in a study testing different metal oxides as well as Ag NPs of different sizes (Karlsson et al. 2014). However, this study showed no evidence for direct DNA interaction leading to stalled replication forks by any of the tested NPs.

Instead, communicator cells showing aerophilic stress were activated chiefly by CuO and NiO NPs. In one in all the few studies claiming a size-dependent interaction with DNA, gold NPs with a distinct particle size of 1.4 nm were shown to interact uniquely with the foremost DNA grooves, which may account for the toxicity of tiny NPs [3,4].

Excluding from the direct DNA interaction, there are quite a lot of other mechanisms which lead to genotoxicity. Probably, the foremost impact associated with genotoxicity is aerophilic stress. For mechanistic functions, one ought to discriminate between the oxidant-generating properties of particles themselves (i.e. acellular) and their ability to stimulate cellular oxidant generation. ROS may result from reactions at the surface of the NPs or via unharness of redox-active transition ions, such as $Fe2^+$, Ag^+, Cu^+, $Mn2^+$, and $Ni2^+$, leading to the production of ROS via the Fenton-type reaction. An example of stimulation of cells to release ROS is an interaction with mitochondria that may affect the electron transport chain or ROS formation via the induction of P450 enzymes. NPs may additionally affect proteins concerned in DNA repair or in inhibitor response, as well as leading to genotoxicity via indirect mechanisms [4].

Another risk is that NPs act with the mitotic spindle equipment, centrioles or their associated proteins, and thereby cause aneugenic effects, that is, loss or gain of chromosomes in daughter cells. As summarised by Sargent et al. (2010), the long thin tubular-shaped CNTs have a striking similarity to cellular microtubules, suggesting a potential to interact with the mitotic spindle because of the motor proteins that separate the chromosomes throughout the organic process. Such disruption of centrosomes and mitotic spindles would result in monopolar, tripolar and quadripolar divisions of chromosomes resulting in aneuploidy. Aneugenic effects for example have also been for CuO and gold NPs. Except for these mechanisms, other indirect mechanisms have been suggested [5].

SECONDARY (INFLAMMATION-INDUCED) GENOTOXICITY

As a distinction to primary genotoxicity, secondary genotoxicity will be outlined as a genetic harm ensuing from reactive oxygen/nitrogen species (ROS/RNS) (and probably alternative mediators) that square measure generated throughout particle-elicited inflammation from activated phagocytes (macrophages, neutrophils). One important factor for risk assessment is that secondary genotoxicity is considered to involve a threshold [5].

GENOTOXICITY APPRAISAL OF NANOMATERIALS

The regulatory test battery for genotoxicity consists of:

- an *in-vitro* test for mutations in bacteria, and
- an *in-vitro* test for cytogenetic effects, micronuclei or mutations in mammalian cells.

In some cases, *in-vivo* test is also needed, for example, micronuclei in erythrocytes or chromosomal aberrations in bone marrow cells.

Test methods that are frequently incorporated *in vitro* are:

- *OECD 471* – Bacterial reverse mutation test
- *OECD 473* – *In-vitro* mammalian chromosome aberration test
- *OECD 487* – *In-Vitro* mammalian cell micronucleus test
- *OECD 476* – *In-vitro* mammalian cell gene mutation test

The scientific literature suggests that the most commonly used assays for evaluating the genotoxicity of NPs are the Comet assay and the analysis of micronuclei *in vitro*.

In the next section, the applicability of these different methods has been discussed [6].

Genotoxicity appraisal of NMs is through the following assays/tests;

- Comet assay
- Micronucleus assay
- Chromosome aberrations
- Bacterial mutagenicity test
- Mammalian mutagenicity tests
- Cell transformation
- High-throughput methods

COMET ASSAY

This assay is the most used assay used for assessing the genotoxicity of NPs. It was first described in 1984 by two Swedish researchers, Östling and Johanson. It is also known as single-cell gel electrophoresis.

A few years later an alkaline version (where pH > 13) of the method was introduced, which remains the most widely used assay. It enables the detection of single-strand breaks, which are directly produced or associated with incomplete excision repair, as well as alkali-labile sites [7–11].

The latter includes a basic site that arises from the trouncing of a dented base from the sugar in the DNA backbone. Such sites can arise instinctively due to damage in the bases or the sugars or as intermediates during base excision repair.

ASSAY METHODOLOGY

- The assay method initiates with embedding the cells (subsequent exposure) in low melting point agarose gel on a microscopic slide.
- After the gels are solidified, the microscopic slides are placed in a lysis solution containing Triton X-100 (needed to break down the membranes) and high concentration of a salt solution (2.5 M NaCl), which is responsible for the removal of histones and other soluble proteins.
- The supercoiled DNA is attached to a nuclear matrix to create a structure called "nucleoid." These slides are subsequently incubated in an alkaline electrophoresis buffer.

- This leads to the unwinding of DNA. Electrophoresis is performed under the same alkaline conditions for approximately 20–30 min at 0.7–1.15 V/cm.
- The electric field enables the negatively charged damaged DNA to migrate toward the anode. An image that looks like a comet with a head and a tail is generated. Thus, with more strand breaks in the DNA, the amount of DNA will increase in the tail.
- After neutralisation, the slides are stained to analyse the comets by fluorescence microscopy. Image analysis calculates various parameters for each comet.
- The most often used parameters are:
  - Tail length
  - Percentage DNA in the tail (% tail DNA)
  - Tail moment (Tail length × total tail intensity).

In general, percentage tail DNA is considered and regarded as an easier option to interpret and is more useful.

The Comet assay can also be tailored to allow the explicit detection of oxidatively damaged DNA. Following the lysis step, a lesion-specific endonuclease is added which removes the damaged base, thereby creating an abasic site that, via subsequent alkaline treatment or by the lyase activity of the enzyme, is transformed to a strand break [7–11].

The commonly used enzymes are enlisted below:

- Formamidopyrimidine DNA glycosylase (FPG)
- Endonuclease III (Endo III)

The disparity in tail intensity between cells treated with enzymes and untreated cells (net FPG or Endo III sites) indicates the amount of oxidatively damaged DNA.

CAN THE METHOD BE APPLIED TO NANOMATERIALS?

- Some issues for interactions of NPs within the Comet assay have been raised from the observation that NPs can be observed *in vitro* in the "Comet Head" on subsequent exposure.
- These imply that NPs may be present during the performance of the assay, which leads to the query that such NPs tempt additional breaks in "naked DNA" during the assay. Thus, DNA damaging particles present during the assay cause additional DNA damage
- Overall, it is clear that there is a risk of overestimating the damage to DNA when there is *in-vitro* usage of NPs [12].

MICRONUCLEUS ASSAY

- It is the second most used assay for appraising the genotoxicity of NPs. Micronuclei mainly originates from acentric chromosomes and chromatid fragments that fall short to be incorporated in the daughter nuclei at the closing stages of telophase during mitosis cell division because of the spindle defects during the separation process in anaphase.

- Micronuclei-containing chromosomes or chromatid fragments are enclosed by a nuclear membrane which shows similar morphology to nuclei after predictable nuclear staining, with the exception for smaller size, which is 1/18 and 1/3 of that of the main nucleus.
- The micronuclei assay detects both chromosomal breakages (clastogenicity; e.g. induced by ROS), as well as aneuploidogenic effects caused by physical disturbance of spindle/mitotic apparatus. The OECD test guideline no. 487 mentions the *in-vitro* mammalian cell micronucleus test. The *in vivo* mammalian erythrocyte micronucleus test is as per the OECD test guideline 474 (OECD 487, OECD 474) [13].

CHROMOSOME ABERRATIONS

- The OECD guidelines TG 473 defines *in-vitro* mammalian chromosome aberration test and TG 475 outlines the *in-vivo* mammalian bone marrow chromosomal aberration test.
- The chromosome aberration test identifies agents that cause notable translocations, which are implicated in the aetiology of various human genetic diseases and cancers. These are chromatid-type or chromosome-type breakage and exchanges, endoreduplications, dicentric chromosome formation and other abnormal chromosomes.
- For *in-vitro* and *in-vivo* testing, cell cultures or animals (generally rodents) are exposed to the test substance and treated with a metaphase-arresting substance (e.g. colcemid), which results in accumulation of metaphase cells. Chromosome preparations are then prepared from cultured cells or bone marrow cells and metaphase cells and are microscopically analysed [14].

BACTERIAL MUTAGENICITY TEST

- The Ames test (Bacterial reverse mutation, OECD, 1997) is based on the stimulation of back- mutations in a defective histidine gene. The delay in this mutation facilitates the bacterium to synthesise histidine which forms a visible colony when plated in a minimal histidine medium [14].
- In short, a typical Ames test reveals the bacterial strain first to a test agent and then by placing the exposed bacteria in Petri dish containing agar without histidine. After incubating the bacteria grown are counted. This reflects the bacteria that have suffered a reverse mutation. A comparison is made with the number of bacteria that have undergone reverse mutations when they have not been exposed to the agent [14].
- The agent is caused to be mutagenic if it causes too many reverse mutations above measured as spontaneous.

MAMMALIAN MUTAGENICITY TESTS

- OECD guidelines are available for *in-vitro* mutation assays that enable the detection of forward mutations in reporter genes.

- *In-vitro* mammalian cell gene mutation tests use either the hypoxanthine-guanine phosphoribosyl transferase (*Hprt* in rodent cells and *HPRT* in human cells) or the xanthine- guanine phosphoribosyl transferase transgene (*gpt*) (*XPRT*) genes to detect different spectra of mutagenic events.
- The *HPRT* test detects mutational events such as the base pair substitutions, frameshift mutations, small deletions and insertions as well. The *XPRT* assay allows the detection of mutations resulting from large deletions and mitotic recombination due to the autosomal location of the *gpt* transgene (OECD 476).
- Cells are exposed to the test NPs and then sub-cultured for an ample amount of time to resolve cytotoxicity and let phenotypic expression be before mutant selection using the purine analogue 6-thioguanine. Mutant frequency is calculated based on the number of mutant colonies rectified by the cytotoxicity at the time of mutant selection [15].

CELL TRANSFORMATION

- The assessment approach of *in-vivo* carcinogenicity is based on the OECD guideline TG 451 (OECD 451).
- The carcinogenic properties of a test substance are recognised by the escalating incidence of neoplastic histopathological findings following 2 years of oral, dermal or inhalation administration in rodent species.
- Numerous *in-vitro* methods have been developed to evaluate the carcinogenic potential of a test substance.
- Among these, the most used in evaluating the carcinogenic potential of NPs are:
  - Cell transformation assays (CTA)
  - Soft agar colony-forming assay
- The CTA uses established cell lines to evaluate the ability of test compound to induce morphological neoplastic transformation of treated cells evaluated by their ability to form morphologically transformed colonies.
- The soft agar colony-forming assay measures cell anchorage and independent growth *in vitro* by manual counting of colonies in semi-solid culture media. The assay has been used in studies testing NMs [16].

HIGH-THROUGHPUT METHODS

The assays used above are regarded as "low throughput" and therefore there is a need for assays that facilitate additional high-throughput screening.

Several such assays and their applicability for testing NMs are described below:

- Comet chip assay and flow cytometry scoring of micronuclei.
- Fluorimetric detection of alkaline DNA unwinding assay.
- Use of reporter cell lines is a striking approach to make enable high-throughput analyses of an array of NMs which fluoresce upon commencement of signalling pathways and to get insight into the diverse mechanisms

of genotoxicity. For this, a combination of an assortment of reporter cell lines would be required.

- ToxTracker assay was developed by Hendriks and co-workers in 2011. The assay consists of a panel of six mouse embryonic stem (mES) cell lines that each contains a different GFP-tagged reporter for a distinct cellular signalling pathway. The assay was adapted to a 96-well plate format enabling high-throughput screening.
- There are several advantages associated with the ToxTracker assay in NMs genotoxicity studies. The mES cells that are used in the ToxTracker assay are untransformed, dexterous in all major DNA damage and cellular stress response pathways and have been shown to efficiently engulf NPs. The ToxTracker reporter cell assay can be applied as a rapid mechanism-based tool for assessing the potential genotoxic effects of NPs [17].

CONCLUSION

In this chapter, the most common assays for testing the genotoxicity of NMs have been described and discussed. In general, the assays used had some interference and drawbacks, which were identified. For the Comet assay, a risk for overestimation of the DNA damage has been suggested when high concentrations of reactive NPs are tested *in vitro* because of the additional damage caused during the assay performance. Most likely, the NPs that cause additional damage also cause real damage, and thus, the risk for false positives seems rather small.

Numerous test schemes and processes have been utilised and employed to review and evaluate the genotoxicity of NMs, with almost a comparable echelon of several outcomes and upshots.

Thus, a small amount of wrapping up on NM genotoxicity can be prepared, regardless of considerable, significant and extensive body of work. Through this chapter an attempt has been made to assess and evaluate the literature, with an observation on the condition regarding commendations and counsel on corroborated and authenticated methods and systems for assessing genotoxicity of NMs.

A massive number of topics and concerns have been documented in this analysis, which is inclusive of an extensive distinction in the physical and chemical properties of NMs, unswerving NM categorisation and depiction in the test medium, numerous test systems over and over again deteriorating to meet the OECD standards, the obscurity of applying NMs to biological systems which are inclusive of its uptake, meddling and prying of NM with the test endpoint, impending and prospective disparity in the systemic allocation *in vivo*, and be deficient in a definitive mechanism of action.

Based on the current data, it is clear that NM genotoxicity responses are less significant than pragmatic conventional DNA damaging agents, which depends on the genotoxicity stimulated via a derivative consequence to a certain extent than an upshot of undeviating DNA interaction.

As an approach to the fore, the following recommendations can be proposed:

- The utility of cautiously defined NMs is inclusive of the categorisation in the test medium.

- An appraisal and evaluation of uptake and allocation within cells and *in-vivo* systems.
- Carefully chosen dose assortment to keep away from object interrelated to system overwork.
- A custom-made test battery that is inclusive of genotoxicity testing in *in-vitro* mammalian mutagenicity and chromosomal damage assays, combined with assay variation, as described within.
- Observing investigations and cell systems as illustrated and expressed in the OECD TG.
- A superior and better effort on the considerate mechanisms.

REFERENCES

1. Arai, Y., Miyayama, T., Hirano, S. 2015. Difference in the toxicity mechanism between ion and nanoparticle forms of silver in the mouse lung and in macrophages. *Toxicology* 3:84–92.
2. Armand, L., Tarantini, A., Beal, D., et al. 2016. Long term exposure of A549 cells to titanium dioxide nano particles induces DNA damage and sensitizes cells towards genotoxic agents. *Nanotoxicology* 10:913–923.
3. Asare, N., Duale, N., Slagsvold, H., et al. 2016. Genotoxity and gene expression modulation of silver and titanium dioxide nano particles in mice. *Nanotoxicology* 10:312–21.
4. AshaRani, P., Low Kah Mun, G., Hande, M., et al. 2009. Cytotoxicity and genotoxicity of silver nano particles in human cells. *ACS Nano* 3:279–290.
5. Balansky, R, Longobardi M, Ganchev G, et al. 2013. Transplacental clastogenic and epigenetic effects of gold nanoparticles in mice. *Mutat Res* 751:42–48.
6. Braakhuis, H., Kloet, S., Kezic, S., et al. 2015. Progress and future of in vitro models to study translocation of nano particles. *Arch Toxicol* 89:1469–1495.
7. Browning, C., The, T., Mason, M. 2014. Titanium dioxide nano particles are not cytotoxic or clastogenic in human skin cells. *J Environ Anal Toxicol* 4:239.
8. Cardoso, E., Londero, E., Ferreira, G., et al. 2014. Gold nano particles induce DNA damage in the blood and liver of rats. *J Nanoparticle Res* 16:8.
9. Cardoso, E., Rezin, G., Zanoni, E., et al. 2014. Acute and chronic administration of gold nano particles cause DNA damage in the cerebral cortex of adult rats. *Mutat Res* 766:25.
10. Catalán, J., Järventaus, H., Vippola, M., et al. 2012. Induction of chromosomal aberrations by carbon nano tubes and titanium dioxide nano particles in human lymphocytes in vitro. *Nanotoxicology* 6:825–836.
11. Catalan, J., Siivola, K., Nymark, P., et al. 2016. In vitro and in vivo genotoxic effects of straight versus tangled multi-walled carbon nanotubes. *Nanotoxicology* 10:794–806.
12. Chen, Z., Wang, Y., Ba, T., et al. 2014. Genotoxic evaluation of titanium dioxide nano particles in vivo and in vitro. *Toxicol Lett* 226:314–319.
13. Corvi, R., Aardema, M.J., Gribaldo, L., et al. 2012. ECVAM prevalidation study on in vitro cell transformation assays: General outline and conclusions of the study. *Mutat Res* 744:12–19.
14. Decan, N., Wu, D., Williams, A., et al. 2016. Characterization of *in vitro* genotoxic, cytotoxic and transcriptomic responses following exposures to amorphous silica of different sizes. *Mutat Res Genet Toxicol Environ Mutagen* 796:8–22.
15. Di Bucchianico, S., Fabbrizi, M., Cirillo, S., et al. 2014. Aneuploidogenic effects and DNA oxidation induced *in vitro* by differently sized gold nanoparticles. *Int J Nanomed* 2191–2204.

16. Di Bucchianico, S., Migliore, L., Marsili, P., et al. 2015. Cyto and genotoxicity of Gold nano particles obtained by laser ablation in A549 lung adenocarcinoma cells. *J Nanopart Res* 17:213.

17. Suzuki, T., Miura, N., Hojo, R., et al. 2016. Genotoxicity assessment of intravenously injected titanium dioxide nano particles in gpt delta transgenic mice. *Mutat Res Genet Toxicol Environ Mutagen* 802:30–37.

10 Genotoxicity Evaluation in Phytopharmaceuticals
Perspective Review

INTRODUCTION

Since ancient times, people have believed that herbal products or herbal medicinal products (HMP) do not cause any risk to health. However, the term natural assigned to any product does not assure any health risk. Herbal preparations or HMPs have some characteristic features similar to other synthetic products and possess some differentiating parameters which make them unique [1–3]. Some of these characteristics are mentioned below:

- They are made of natural substances.
- They contain active constituents in the form of a complex mixture in varying amount.
- Their composition varies according to the geographical location, harvesting method and climatic condition.
- Most bioactive constituents are unknown, which may contribute to hidden danger or hazards to humans and society.
- Their safety is of critical concern when used for long times and in high doses.
- There is no specific legislation to regulate HMPs [4–7].

Because of a lack of standardization and low quality, adverse effects are the major concern raised for herbal medicines. Although phytoactive compounds or herbal products are safe and effective in a small dose, high-dose and long-term use of these plants-based products may become dangerous and lead to a certain toxicity. Unfortunately, long-term exposure to herbal products are associated with systemic toxicity and genotoxicity, thereby resulting in an increase in the morbidity and mortality rates.

Overdose is higher in the case of herbal preparations than conventional medicines due to product variability. This may be because of the lack of adequate regulatory guidelines for safety and use of herbals [8]. Most plants produce toxic secondary metabolites as natural defence from adverse conditions which are not distinguished from their therapeutically active compound. Despite the extensive faith and use of herbal products by human civilisation, only 10%–12% of phytomedicines in the world are truly standardized to known active components, and strict quality control measures are not always diligently followed [9,10].

The belief from ancient times that natural/herbal medicines are much safer than synthetic drugs has caused exceptional growth in human exposure to natural products as bioactive compounds, phytotherapeutic agents and phytopharmaceutical products. This information has led the way to the resurrection of methodical significance in their biological effects.

Till today, most countries lack a universal regulatory system for ensuring the safety, efficacy and toxicity of herbals, and they had not been sufficiently investigated analytically or toxicologically. Scientific research has shown and proven that several plants used as medicinal products in traditional and folk medicine are potentially toxic, mutagenic, genotoxic and carcinogenic. Assessment of the potential genotoxicity of such HMPs is indeed an important issue as genetic damage may lead to critical mutations causing an increased risk of cancer and other diseases.

NEED OF TOXICITY STUDY FOR HERBAL MEDICINAL PRODUCTS

Toxicity testing can reveal some of the risks that may be associated with the use of herbs, especially in sensitive populations, thus avoiding potential harmful effects when used as medicine [11]. The basic aim behind the assessment of toxicity in any herbal medicine is to identify adverse effects and to determine the limits of exposure at which such effects occur. For the most part, two factors are well thought out in evaluating the safety of any herbal drug: the nature and implication of the undesirable consequence and the exposure level where the upshot is scrutinised. Toxicity testing in herbals deals with the detection of toxic plant extracts or toxic compounds derived from them during pre-clinical and clinical stages of drug discovery and development from natural plant sources. This will help in the identification of toxins which can be removed, discarded or modified during the process. Such a process will generate an opportunity for an extensive evaluation of safer, promising alternatives. Modifications could be done through dose reduction, formulation into suitable dosage form other than existing ones, changes or alteration in a chemical or functional group or structural adjustments in a phytoactive compound, all of which may improve their tolerability [12–17].

GENOTOXICITY IN HMPs

About 300 years ago, only a few synthetic medicines were available; people used species of higher plants as their main source of medicines. Drug discovery was based on trial and error, which related the cause-and effect relationship to use plants or animals or their parts and the desired result. Either the whole plant or its selected parts like roots, leaves, bark, stem, seeds, fruits, etc. were usually selected. In the case of animals, organs and glands like liver, skin, nails, thorns, etc. were used for therapeutic purposes.

HMPs contain numerous biologically active phytoconstituents or bioactive compounds, which possess pharmacological properties and elicit therapeutic action. Furthermore, they can cause acute or chronic harm including DNA damage. Continuous interaction of human body cells to physiological and external parameters and processes may cause cytotoxic, genotoxic and oxidative damage.

Genotoxicity is defined as a chemically induced mutation or alteration of the structure and/or segregation of genetic material. Most plants and herbs harbour phytochemicals or bioactive constituents that either directly react with DNA or may disturb cellular homeostasis, cell cycle and genome maintenance mechanisms, all of which may contribute to the development of genotoxicity, carcinogenicity or co-carcinogenicity. Genotoxicity studies are a key step for risk assessment during drug development for protecting human health because various genotoxic compounds can cause DNA damage, such as cross-links, adducts and cleavage [18–20]. The entire plant or various parts such as bark, stem, roots, leaves, flowers, fruits and seeds, as well as some by-products such as essential oils, volatile substances and other compounds derived from the secondary metabolism of plants possess the therapeutic activity and elicit a pharmacological response. These parts depending on their use should be assessed for genotoxicity studies. The mitotic index reflects the frequency of mitotic cells and hence the cytotoxicity of crude extracts treatments are analysed. Genotoxicity is a special area in toxicities as it is often the most difficult to detect [21,22].

REGULATORY ASPECTS FOR GENOTOXICITY IN HMPs

Earlier, there was no specific protocol or guidelines for estimation of genotoxicity in herbals. Recently, a guide has been drafted by the European Medicines Agency (EMA) for the assessment of genotoxicity of herbal preparations. The Korean Ministry of Food and Drug Safety (MFDS) has developed guidelines (No. 2015-82, 2015) that require genotoxicity testing before introducing any herb-derived new drug [23–25].

Principles (guidelines) for genotoxicity investigation of pharmaceuticals have been instituted by regulatory authorities such as OECD, ICH and EMEA. In case of medicinal products, genotoxicity testing involves a battery of tests, where prokaryotic and eukaryotic systems are examined through *in-vitro* and *in-vivo* experimental setups with and without metabolic activation.

A stepwise procedure for genotoxicity testing in HMPs was established by the HMPC in the form of guideline on non-clinical documentation for HMP in applications for marketing authorisation and applications for simplified registration [26–29].

Following regulatory agencies are working towards genotoxicity assessment of HMPs:

- ICH – International Conference on Harmonization was established in 1990.
- OECD – Organisation for Economic Cooperation and Development was established 1961.
- Schedule Y – From the Drug and Cosmetic Act was enacted in 1940.

Genotoxicity testing was first published in 1987. The global updates for genotoxicity testing were published in 1997, 2013, 2014, 2015 and 2016, as shown in Table 10.1.

No scientific rationale is available to assume that any plant, its parts and or derived products, including those of long-standing popular use, are intrinsically safe and/or beneficial; compared to conventional medicines, they would require fewer and simpler pre-clinical and or clinical studies. For coherence in drug regulations, all medicines, regardless of their origin and development, should meet equally rigorous safety, efficacy and toxicity standards for marketing authorisation.

TABLE 10.1

Revised and Updated Guidelines for Genetic Toxicity Assessment of HMPs

Guidelines	Test
Genotoxicity Test [TG] According to OECD Guidelines	
TG 471	Related to the Ames Test
TG 487	Related to the *in-vitro* mammalian cell micronucleus test
TG 474	Related to the mammalian erythrocyte micronucleus test
TG 473	Related to the *in-vitro* mammalian chromosomal abrasion test
TG 475	Related to the mammalian bone marrow chromosomal aberration test
TG 483	Related to the draft for mammalian spermatogonial chromosomal aberration test
According to ICH Guidelines	
S2A	Related to the guidelines on explicit facets of regulatory test for pharmaceuticals
S2B	Related to the guidelines on genetic toxicity; a standard battery for the testing of pharmaceuticals
M3	Related to the point in time of pre-clinical studies in relation to clinical trials
According to EMA	
EMA	Related to the evaluation of genetic toxicity of herbal substances and preparations

GENOTOXICITY ASSESSMENT

During research and development of conventional available synthetic drugs, the genotoxic potential is assessed through a battery of *in-vitro* and *in-vivo* short-term tests. A positive result in short-term tests necessitates long-term testing, which is both costly and time-consuming. A waiver of long-term carcinogenicity studies for marketing approval can be obtained if results of short-term genotoxicity tests are negative and the drug is intended to be used continuously for less than 3 months, or intermittently for less than 6 months. The short-term tests for genotoxicity are typically used to identify potential mutagens and carcinogens, however, the same methods can also be used to identify anti-genotoxic agents.

Long-term carcinogenicity studies have seldom been performed with HMP, and reports of their genotoxic potential are scant as well. Owing to their limited duration, clinical trials do not shed light on the carcinogenic potential of drugs. In principle, evidence regarding the carcinogenicity of traditional herbal medicines could be obtained from observational epidemiological studies. No single validated test can provide information regarding genotoxicity, as well as its critical endpoints which are mutation-induced.

This involves dealing with a battery of tests to determine the genotoxic potential of a compound. Because of the diversity of endpoints, genotoxicity cannot be assessed in a single method. Genotoxicity assays aim to detect compounds that can induce genetic damage by various mechanisms [30]. The major challenge in genotoxicity testing resides in the development of methods that can reliably and sensibly detect either such a vast array of damages or a general cellular response to genotoxic insult. Methods for genotoxicity assessment include *in-silico* prediction method, structure alert method, *in-vitro* method and *in vivo* method [31] (Figure 10.1).

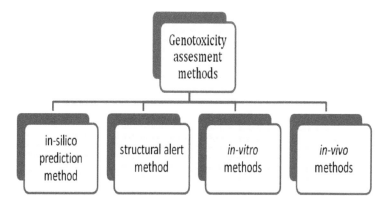

FIGURE 10.1 Methods for genotoxicity assessment.

In-Silico Prediction Method

This method focuses on predicting the biological activities of any molecule based on its physicochemical properties. These prognostic methods work on data interpretation through the use of computational tools and mathematical calculation [7,32]. The analysis of experimental and predicted data is done using computer-based models.

The models are generally classified as:

a. Three-dimensional – docking model: This model is three-dimensional and works on computational docking of DNA for identification of molecules which are capable of non-covalent interaction.
b. QSAR model: This quantitative structure–activity relationship model uses electro-topological descriptors rather than chemical structure to predict mutagenic reactivity with DNA, for example, TOPKAT.
c. Rule-based expert model: This system is used to estimate the presence of a DNA-reactive moiety in any given molecule, for example, DEREK [33–35].

Benefits

The unique advantage of this method is the 3R policy – replacing, reducing and refining the use of animals. Other advantages are its low-cost, rapid, high reproducibility, low/no compound synthesis requirements and constant optimisation.

Limitations

- Endpoint detection is inappropriate.
- Lack of factual toxicity data.
- Models are applicable for pre-decided domains.
- Complex mixtures like herbal extracts are difficult to estimate due to the presence of multiple constituents in a single extract, and it is limited to the detection of known or new structural alerts for genotoxicity.
- Compounds that are reported and responsible for known pharmacological action could be elucidated by this method.

STRUCTURE ALERT METHOD OR TOXICOPHORES

These are the molecules which are associated with toxicity. According to the name, their presence in a compound alerts the investigator regarding the presence of potential toxicity. Through this method, some of the well-known genotoxic compounds were identified from plant constituents belonging to alkaloids, flavonoids, essential oils, etc. Some of them are 1-2 unsaturated pyrrolizidine ester from plants of family Boraginaceae, Asteraceae and Fabaceae; few aristolochic acids and nitro-polyaromatic compounds responsible for terminal nephropathies observed upon intoxication by many Aristolochia species. In addition, allylalkoxybenzenes like eugenol, methyleugenol, estragole, safrole and asarone were potentially genotoxic components from some essential oils [36,37].

IN-VITRO METHODS

These experiments are carried out in a controlled environment outside of a living organism. The term *in vitro* is derived from the Latin word meaning in the glass. These methods are based on the use of prokaryotic and/or eukaryotic cells and tissue cultures. However, because human cells predict human toxicity much better, human cells are now recommended for the study of genotoxicity [38].

Benefits

- Relatively inexpensive
- Cost-effective
- Easy to conduct
- Easy to operate
- Does not require animals for the study

In-vitro methods or assays are just an indicator of the genotoxicity in any compound; the results obtained reveal confirmation followed by *in-vivo* studies.

Limitations

Oversensitivity and low specificity are common issues; requires additional efforts like supplementation with exogenous metabolic activation enzymes to simulate mammalian metabolism. Uses of cell lines is not relevant to predict genotoxic endpoints at target organs. Most cell lines are deficient in DNA repair, p53 function or metabolic competency, and many derive from malignancies.

IN-VIVO METHODS

The term *in vivo* refers to experiments based on a whole, living organism – as opposed to a partial or dead organism – and consists of either animal studies or clinical trials. Because of evidence-based ethical reasons, it is necessary to conduct genotoxicity studies in animals. *In-vivo* methods are done only after the completion of *in-vitro* investigations. *In-vivo* studies also help to overcome their limitations associated with *in-vitro* methods [39,40].

Benefits

In-vivo study leads to complete investigation of pharmacokinetic parameters and factors that influence the outcomes of genotoxicity assessment, which allows better extrapolation of potential noxious effects to humans.

Limitations

In-vivo studies are chronic, time-consuming and costly. As the metabolic activity of drugs or compounds significantly varies among mammals, both negative and positive data may not be transferrable to humans. *In-vivo* tests, such as the bone marrow micronucleus test, are relatively insensitive, therefore, the established *in-vitro* genotoxicity tests are still considered first-line tests as they are sensitive enough to detect a great majority of genotoxins. For drugs and/or compounds with poor systemic absorption like aluminium-based antacids and radioimaging agents; *in-vivo* test on bone marrow, blood or liver are unable to provide exact and useful information; hence, they are rarely used [41].

According to HMPC, the stepwise testing process for HMP involves a battery of genotoxicity tests which are adapted from EMA. It is well known that HMPs contains a large number of characteristic features that differentiate them from other medicinal products and thus require specific guidance. Herbal products are complex mixtures containing a large number of constituents that are sometimes present in highly variable amounts. The complete composition of a preparation is often unknown, and the composition varies with many parameters depending upon geographical origin, method of preparation, harvesting conditions, contamination, adulteration, etc.

All these could invalidate the previously available data regarding genotoxicity [42,43]. Nevertheless, HMPs are framed by similar regulations as for other medicinal products for human use; as with other medicinal products, signals of adverse effects could arise occasionally through pharmacovigilance. The detailed step-by-step procedure is shown in Figure 10.2.

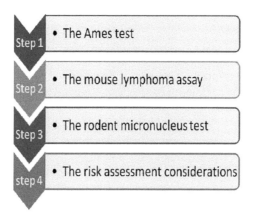

FIGURE 10.2 Four steps of genotoxicity testing of HMPs according to HMPC.

STEP 1: BACTERIAL REVERSE MUTATION TEST/THE AMES TEST

Ames test, also known as the bacterial reverse mutation test, should always be performed and interpreted in conformity with the existing OECD and EU guidelines. This test reveals the mutagenic potential of any substance in a prokaryote organism and whether the reactive metabolite is a product of metabolic activation by mammalian enzymes. To conduct this study, a set of different *Salmonella typhimurium* strains with various mutations present in an amino acid-synthesising gene is incubated under selected pressure.

Mutations occurring in the non-functional gene restore the capability of bacteria to synthesize the specific and essential amino acid also referred to as revertants. These are counted and correlated with the mutagenic potential of a substance. The first stage, that is, Ames test with *S. typhimurium* does not always lead to the identification of genotoxic compounds and may also lead to false results in the form of positive data [35,38].

Thus, there is a need to proceed with more reliable tests like the mouse micronucleus test and mouse lymphoma assay (MLA) to confirm the results. If the result of step 1 is negative, no further genotoxicity testing is required. If the result of step 1 is positive, the presence of acknowledged genotoxic compounds not known to be carcinogenic can tentatively explain the mutation. However, the absence of such genotoxic compounds implies that the herbal product has to be studied in a step 2 test.

Equivocal Test Result

For compounds with weak results of genotoxicity or non-consistent regarding the usual positive response in the test, special considerations are required. The first option is to repeat the test to reveal whether the test outcome is the same as in the original experiment [29,37].

Under such conditions, proper assessment and survey should be done for at least some of the critical parameters:

- Whether the response is dose-dependent?
- Does it exhibit unusual or irregular features concerning the concentration?
- Are there indications that the preparation affects the growth of test organisms, thus preventing the detection of genotoxic constituents?

Based on these questions, the final decision should be made by transparent consideration of the outcomes in the light of test material and test conditions. Equivocal test results require special considerations, and a repetition of the experiment should generally be envisaged.

STEP 2: THE MOUSE LYMPHOMA ASSAY OR OTHER MAMMALIAN CELL ASSAY

Similar to the Ames test, the MLA or other mammalian cell assay should also be performed and interpreted in conformity with the existing OECD and EU guidelines.

Briefly, mouse lymphoma cells, that is, L5178Y, present in culture are exposed to a compound or a preparation, as well as mutants in the thymidine kinase gene. These are detected by their resistance to the cytotoxic pyrimidine analogue trifluorothymidine. This assay may confirm or prove false-positive findings in the Ames test.

In case of HMP, it also provides information on its ability to cause chromosomal damage. If other mammalian cell assays are used for genotoxicity tests, their use has to be justified.

- In the case of a negative result, no further testing is required.
- In the case of a positive result, the relevance of the finding should be thoroughly assessed as it is known that the MLA is associated with false positives. If the test is unequivocally positive (gene mutation or chromosomal damage), it is advisable to proceed to step 3. If the herbal preparation is known to contain a compound with chromosomal damaging properties, it may be advisable to perform the *in-vitro* micronucleus test in mammalian cells in culture.

STEP 3: THE RODENT MICRONUCLEUS TEST

The rodent micronucleus test should also be performed and interpreted in conformity with the existing OECD and EU guidelines. The micronucleus test was first reported in 1970 by Boller and Schmid and was later used by Heddle in 1977. The micronucleus is a supplementary nucleus which is separated from the core nucleus of a cell for the duration of its division. It is composed of whole chromosomes or chromosomal fragments that remain from other chromosomes after the completion of mitosis [27,29].

The micronuclei result from structural changes in the chromosome, spontaneous or experimentally induced, or even by cell fusion errors. However, these micronuclei are excluded from the new nuclei renewed at telophase. For a short time, mice or rats are administered with the compound or preparation prepared in an apposite vehicle and via the suitable route of administration. The proportion of micronuclei in the bone marrow and/or peripheral blood cells can identify agents causing structural and numerical chromosomal changes [43].

- In case a negative result is obtained, no further testing is required.
- In case a positive result is obtained, it is always advisable to proceed to step 4.

STEP 4: THE RISK ASSESSMENT CONSIDERATIONS

Because of the complexity of HMP, no single specific approach has been recommended for risk assessment. A risk assessment through the threshold of toxicological concern (TTC) approach is possible whenever a herbal preparation contains an identifiable genotoxic compound that presents a demonstrated threshold mechanism; permissible exposure levels – without appreciable risk of genotoxicity – can be established according to the usual no observable effects level method.

However, as herbal preparations are complex mixtures with partially unidentified components, it is quite possible that the compound(s) responsible for genotoxicity is and are still not identified at the end of the testing protocol. Thus, the usual procedure for toxicity testing and risk assessment of mixtures should consist of isolating and identifying various major constituents and testing them individually – which is a time-consuming, costly and probably unrealistic approach for herbal medicines.

This HMPC stepwise testing process for HMP effectively defines the Ames test as the primary endpoint which, if negative, accepts the drug as probably non-genotoxic. This is not entirely satisfying, however, and has been greatly debated, indeed (1) the Ames test does not detect every genotoxic insult; and (2) because some common compounds, including flavonoids, yield very positive Ames tests but are not carcinogens, they may effectively mask the genotoxic effect of real carcinogens. A TTC value of 1.5 μg/day ingestion of a genotoxic impurity is considered to be connected with an adequate risk (surfeit cancer risk of <1 in 100,000 over a lifetime) for most pharmaceuticals. From this threshold value, a permitted level in the active substance can be calculated based on the expected daily dose. Higher limits may be justified under certain conditions such as short-term exposure periods. The same approach might be considered for genotoxic constituents in herbal substances/preparations if sufficiently justified by the applicant. Moreover, higher limits may be applied when the applicant submits additional data and a toxicologically plausible argumentation for the required justification [44].

The schematic stepwise decision tree for genotoxicity testing of HMPs is shown in Figure 10.3.

APPLICATION OF THE BRACKETING AND MATRIXING

Bracketing and matrixing involve extrapolation of the results obtained with a specific preparation to closely related preparations such as extracts prepared with alcohol and water mixtures of different, but similar concentrations. Using such an approach

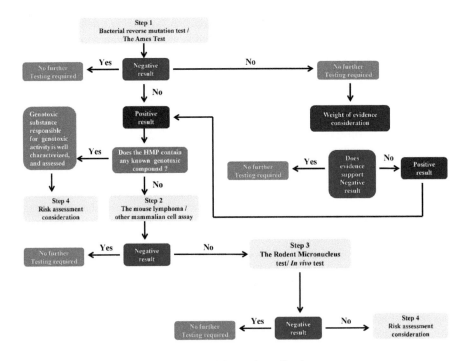

FIGURE 10.3   Decision tree for genotoxicity testing of herbs.

to the test materials implies that a representative range of materials is tested rather than requiring individual manufacturers to undertake their testing on all their specific preparations. This reduced test design assumes that the genotoxic potential of any intermediate preparation is represented by the test results of the extremes tested [45].

For testing, extracts should be characterised according to their individual specifications as provided by the cooperating pharmaceutical companies. Tests should be conducted by GLP-conforming laboratories according to the current guidelines, including those of OECD, ICH and EMA. Validation may also be included for test results by independent testing in two laboratories (Figure 10.4).

ALLIUM CEPA TEST

This test uses the root system of the plant. This method is in use since 1938 (Levan, 1938) and is used for investigating environmental pollution factors, as well as genotoxic and anti-cancer potential of medicinal plants. According to Rank and Nielsen (1994) *A. cepa* test is more sensitive than the Ames test. The mitotic index and replication index are used as indicators of adequate cell proliferation which can be measured by the plant test system *Allium cepa*. This test uses a model that is adequately sensitive to detect most of the substances that cause chromosomal alterations.

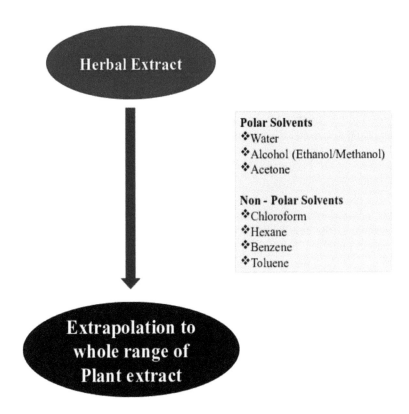

FIGURE 10.4   Bracketing and matrixing approach.

The *Allium cepa* test is an excellent *in-vivo* model where the roots grow in direct contact with the substance of interest (i.e. effluent or intricate medicinal blend being tested which will enable the probable damage to the DNA of eukaryotes to be envisaged). Therefore, the data can be extrapolated for all animal and plant biodiversity. The analysis of chromosomal alterations can be equal to the test of mutagenicity mainly for the detection of structural and numerical alterations. The *A. cepa* test is one of the few unswerving methods for the measurement of smash up in systems that depict the mutagens or probable carcinogens and facilitate the estimation of the consequences of these smash-ups through the surveillance of chromosomal alterations. The main component of *A. cepa* is a vascular plant, which makes it an excellent genetic model for evaluating environmental pollutants, detecting mutagens in different environments and evaluating many genetic endpoints (point mutations to chromosomal alterations).

ADVANTAGES

This test is easy, feasible, cost-effective and less time-consuming. It lasts for 4 days. It provides good chromosomal conditions for the study of chromosomal and cell damage which can be extrapolated on animal cells.

NOVEL APPROACHES FOR GENOTOXICITY TESTING

Omics is, in general, an approach to the cooperative technologies employed to investigate the functions, associations and deeds of the diverse types of molecules that comprise the cells of an organism. Omics studies involve multiple measurements per endpoint to acquire a comprehensive, integrated understanding of biology and to identify various factors simultaneously (e.g. genes, DNA, RNA, proteins and other metabolites) rather than each of those individually.

Some of the newer areas in genomics include:

- Toxicogenomics is the novel branch that deals with the swot up of the interactions among the structure and activity of the genome and the unsympathetic biological consequences of exogenous agents.
- The noxious effects of xenobiotics on biological systems are, by and large, imitated at the cellular level by their brunt on gene expression which is studied under the heading transcriptomics, and on the production of proteins is studied under the head as proteomics, and small metabolites are studied under the branch metabonomics.
- Genetic disparity and appearance signatures can be utilized and employed to screen compounds for associated vulnerability, to weigh the cellular rejoinders to a choice of dosages, to categorise toxicants on the starting point of mechanisms of action, to keep an eye on the revelation of folks to the toxicants and to envisage the individual unpredictability in compassion to the toxicants.
- Toxicogenomics effactually consents to the comprehending of dose-response relationships, cross-species extrapolations, quantifications of the exposure, fundamental mechanisms of toxicity and the starting point of personage vulnerability and inclination to particular compounds [46–48].

DISCUSSION AND CONCLUSION

Although not much literature is available for the genotoxicity of herbals, some plants whose genotoxicity has been studied and reported are discussed here. In 2012, Kelber et al. compiled a review on genotoxicity of herbal plants listing the name of plants and highlighting how the results were obtained. In 2014, an experimental study was conducted for the genotoxicity assessment of *Valeriana officinalis* L., radix root. Sponchiado et al. in 2016 compiled a systemic review of medicinal plants for quantitative genotoxicity assay. He discussed and conferred that several methods for genotoxicity assay are available and can be employed for the evaluation of the genetic toxicity potential of medicinal plant extracts. These methods are highly recommended by regulatory agencies.

Based on the findings, to conduct a thorough study regarding the possible genotoxic effects of any medicinal plant, it is important to include bacterial and mammalian tests, with at least one *in-vivo* assay. These tests should be capable of detecting outcomes such as mutation induction, clastogenic and aneugenic effects and structural chromosome abnormalities.

Mélanie Poivre in 2017 summarized the genotoxicity of phytoconstituents isolated from various plants. Some of the contribution in research for genotoxicity of herbals is listed in Table 10.2.

The concept of toxicity study for HMPs is not new. Since ancient times, it was based on the experience of people about how much, when and where to use these herbal products for the treatment of ailments. Because of adulteration and changes in climatic and geographical conditions, it is the need of the hour to explore genotoxicity, carcinogenicity, mutagenicity, adverse effects, side effects, etc. of all traditional herbal plants. Most of the secondary metabolites have been proven to be toxic in high doses.

However, OECD, WHO, EMA and ICH have collaboratively discussed the detailed procedure for ensuring the safety of herbals through *in-vitro* and *in vivo* toxicity study. This will also help in maintaining data and records for future use of herbals.

List of phytoconstituents and medicinal plants along with their families whose genotoxicity data is already reported in the literature is highlighted in Tables 10.3 and 10.4, respectively.

The experience of the allopathic industry suggests that regulatory guidelines and their implementation are necessary to support science and quality of research. Time has to come to accept the same for HMP. The estimation of genetic biomarkers would help determine the potential toxicity of medicinal herbs to regulate medicinal plant consumption, which would be an important measure of public health protection.

Genotoxic events have been known as a crucial step in the initiation of cancer. To assess the risk of cancer, genotoxicity assays including Comet, micronucleus, chromosomal aberration, bacterial reverse and sister chromatid exchange assay can be performed. Compared with *in-vitro* genotoxicity assay, *in vivo* genotoxicity assay has been used to verify *in-vitro* assay results and provide biological significance for certain organs or cell types [49–52].

TABLE 10.2
Genetic Toxicity Study of Some Plants Reported in Literature

S. No.	Plant	Family	Therapeutic Use	Extract Used	Method of Genotoxicity Assessment
	Goji Berry Lyciumbarbarum	Solanaceae	Anti-oxidant and revitalizing potential, superfruit and dietary supplement	Aqueous	Micronucleus test and Comet assay
	Myelophil (Mixture of *Astragali radix* and *Salviae miltiorrhizae radix*)	-	Used to treat problems of Qi and blood-related disorders	Ethanolic	Ames test, i.e. bacterial reverse mutation test, mammalian chromosome aberration test and mammalian erythrocyte micronucleus test
	Elephantopus scaber (Tutup Bumi)	Asteraceae	Anti-microbial, anti-inflammatory and anti-tumour activities	Aqueous and methanolic leaf and root extracts	*Allium cepa* root tip assay
	Glycyrrhiza uralensis	Fabaceae	Fever, liver ailments, dyspepsia, gastric ulcers, asthma, bronchitis, Addison's disease and rheumatoid arthritis and has been used as a laxative, anti-tussive and expectorant	Aqueous and methanolic root extracts	*Allium cepa* root tip assay
	Salvia miltiorrhiza	Lamiaceae		Aqueous and methanolic root extracts	*Allium cepa* root tip assay

TABLE 10.3
List of Phytoconstituents Whose Genotoxicity is Already Reported

S. No.	Phytoconstituent	Family
1	Anthraquinones	Rubiaceae
2	Aristolochic acids I and II	Aristolochiaceae
3	Asarone	Acoraceae
4	Estragole	Asteraceae
5	Heliotrine	Compositae (Asteraceae)
6	Monocrotaline	Leguminosae (Fabaceae)
7	Myristicin	Myristicaceae
8	Retrorsine	Boraginaceae
9	Safrole	Lauraceae, Myristicaceae

TABLE 10.4
List of Plants Whose Genotoxicity Is Already Reported

1	*Allii sativi bulbus*	Amaryllidaceae or Liliaceae
2	*Althaeae radix*	Malvaceae
3	*Berberidis cortex*	Berberidaceae
4	*Betulae folium*	Betulaceae
5	*Catharanthus roseaus*	Apocynaceae
6	*Cardui mariae fructus*	Asteraceae
7	*Carvi fructus*	Apiaceae
8	*Crataegi folium*	Sapindaceae
9	*Crataegi fructus*	Rosaceae
10	*Curcurbitae oleum*	Cucurbitaceae
11	*Cynarae folium*	Compositae
12	*Dulcamarae stipites*	Solanaceae
13	*Ginkgo folium*	Ginkgoaceae
14	*Ginseng radix*	Araliaceae
15	*Harpagophyti radix*	Pedaliacea
16	*Hippocastani semen*	Hippocastanaceae
17	*Hyperici herba*	Hypericaceae
18	*Liquiritiae radix*	Fabaceae
19	*Lupuli flos*	Cannabaceae
20	*Matricariae flos*	Asteraceae
21	*Melissae folium*	Lamiaceae
22	*Passiflorae herba*	Passifloraceae
23	*Pini aetheroleum*	Pinaceae
24	*Primulae radix*	Primulaceae
25	*Rosmarini folium*	Labiatae
26	*Serenoae repentis fructus*	Arecaceae
27	*Thymi herba*	Lamiaceae
28	*Taxus brevifolia*	Taxaceae
29	*Urticae folium*	Urticaceae
30	*Urticae herba*	Urticaceae
31	*Valerianae radix*	Valerianaceae
32	*Visci albi herba*	Loranthaceae

List of Herbal Plants Whose Genotoxicity Data is Available from the Joint Project of Kooperation Phytopharmaka.s

REFERENCES

1. Avani, K., Neeta, S. 2005. A study of the antimicrobial activity of Elephantopus scaber. *Indian J Pharm Sci* 37:127.
2. Barnes, J., Anderson, L., Phillipson, D. 2007. *Herbal Medicines*, 3rd Edition. Press P, Editor, London.

3. Billintona, N., Hastwellb, P., Beerensc, D., Birrella, L., Ellisb, P., Maskelld, S. 2008. Inter laboratory assessment of the Green Screen HC GADD45a-GFP genotoxicity screening assay: An enabling study for independent validation as an alternative method. *Mutat Res* 653:23–33.

4. Booth, E., Rawlinson, P. 2017. Regulatory requirements for genotoxicity assessment of plant protection product active ingredients, impurities, and metabolites. *Environ Mol Mutagen* 58:325–344.

5. Borner, F.U., Schutz, H., Wiedemann, P. 2011. The fragility of omics risk and benefit perceptions. *Toxicol Lett* 201:249–257.

6. Brusick D. 1980. *Principles of Genetic Toxicology*. Plenum Press, New York, NY.

7. Celik, T. 2012. *Potential genotoxic and cytotoxic effects of plant extracts*. A compendium of Essays on Alternative medicine. 233–250.

8. Chan, S. 2003. Determination of aristolochic acids in medicinal plant and herbal product by liquid chromatography electrospray ion trap mass spectrometry. *Talanta* 60:679–685.

9. Chen, C., Chi, C., Chang, K., Liu, T. 1999. Safrole like DNA adducts in oral tissue from oral cancer patients with a betel quid chewing history. *Carcinogenesis* 20:2331–2334.

10. Chen, T., Mei, N., Fu, P.P. 2010. Genotoxicity of pyrrolizidine alkaloids. *J Appl Toxicol* 30:183–196.

11. Combes, R. 2012. In silico methods for toxicity prediction. In: Balls M, Combes R, Bhogal N (eds) *New Technologies for Genotoxicity Testing*. Springer, New York. 96–116.

12. CPMP Note for Guidance on Genotoxicity: A Standard Battery for Genotoxicity Testing of Pharmaceuticals (CPMP/ICH/174/95).

13. CPMP Note for Guidance on Genotoxicity: Guidance on Specific Aspects of Regulatory Genotoxicity Tests for Pharmaceuticals (CPMP/ICH/141/95).

14. Daisy, P., Mathew, S., Suveena, S., Rayan, N.A. 2008. A novel terpenoid from Elephantopus scaber—antibacterial activity on *Staphylococcus aureus*: A substantiate computational approach. *Int J Biomed Sci* 4:196–203.

15. ECVAM. 2015. European Union reference laboratory for alternatives to animal testing.

16. EMA. 2012. EMA/CHMP/ICH/126642/2008. ICH guideline S2 (R1) on genotoxicity testing and data interpretation for pharmaceuticals intended for human use step 5.

17. EMEA. 2008. Guideline on the assessment of genotoxicity of herbal substances/preparations. London [06/02/2014]; 4–11/2.

18. EMEA. 2006b. Committee on Herbal Medicinal Products (HMPC). Guideline on Non Clinical Documentation for Herbal Medicinal Products in Applications for Marketing Authorization (Bibliographical and Mixed Applications) and in Applications for Simplified Registration. EMEA/HMPC/32116/2005, dated Sept 7.

19. EMEA. 2008. Committee on Herbal Medicinal Products (HMPC). Guideline on the Assessment of Genotoxicity of Herbal Substances/Preparations. EMEA/HMPC/107079/2007, dated May 21.

20. EMEA. 2009. Committee on Herbal Medicinal Products (HMPC). Guideline on Selection of Test Materials for Genotoxicity Testing for Traditional Herbal Medicinal Products/Herbal Medicinal Products. EMEA/HMPC/67644/2009, dated Nov 12.

21. FDA. 2012. Guidance for industry S2 (R1) genotoxicity testing and data interpretation for pharmaceuticals intended for human use.

22. Fu, P., Xia, Q., Lin, G., Chou, M. 2012. Genotoxic pyrrolizidine alkaloids – mechanisms leading to DNA adduct formation and tumorigenicity. *Int J Mol Sci* 3:948–964.

23. Gaedcke, F., Kelber, O., Kraft, K., Steinhoff, B., Winterhoff, H. 2009. Assessment of genotoxicity of herbal medicinal preparations according to the guideline EMEA/HMPC/107079/2007: A model project of Kooperation Phytopharmaka, Bonn, Germany. *Planta Med* 75:994.

24. Greene N. 2002. Computer systems for the prediction of toxicity: An update. *Adv Drug Deliv Rev* 54:417–431.
25. Guideline on Selection of Test Materials for Genotoxicity Testing for Traditional Herbal Medicinal Products/Herbal Medicinal Products. EMEA/HMPC/67644/2009.
26. Guideline on the limits of genotoxic impurities (CPMP/SWP/5199/02, EMEA/CHMP/QWP/251344/2006).
27. Hartung, T. 2011. From alternative methods to a new toxicology. *Eur J Pharm Biopharm* 77:338–349.
28. Hoet P, Godderis L. 2013. Genotoxicity and mutagenicity testing: A brief overview of the main tests, pitfalls and regulatory framework. In: BEMS-Belgian Environmental Mutagen Society B (ed) *Toxicology as the Scientific Basis for Management of Chemical Risk*. BEMS, Elewijt.
29. I.C.H. 2011. (ICH) International Conference on Harmonization of Technical Requirements for Registration of Pharmaceuticals for Human Use. ICH Tripartite Guideline S2B. Genotoxicity: A Standard Battery for Genotoxicity Testing of Pharmaceuticals, European Federation of Pharmaceutical Industries' Associations Brussels.
30. ICH. 2014. Guidance for industry S2B genotoxicity: A standard battery for genotoxicity testing of pharmaceuticals. [03/02/2014].
31. International Conference on Harmonization (ICH). 2011. ICH Consensus Guideline S2 (R1) Guidance on Genotoxicity Testing and Data Interpretation for Pharmaceuticals Intended for Human Use. Step 4 version of November 2011.
32. Isbrucker, R., Burdock, G. 2006. Risk and safety assessment on the consumption of Liquorice root (Glycyrrhiza sp.), its extract and powder as a food ingredient, with emphasis on the pharmacology and toxicology of glycyrrhizin. *Regul Toxicol Pharmacol* 46:167–192.
33. Zhang, J., Wider, B., Shang, H., Li, X., Ernst, E. 2012. Quality of herbal medicines: Challenges and solutions. *Complement Ther Med* 20:100–106.
34. Zhou, J., Ouedraogo, M., Qu, F., Duez, P. 2013. Potential genotoxicity of traditional Chinese medicinal plants and phytochemicals: An overview. *Phytother Res* 27:1745–1755.
35. Kelber, O., Steinhoff, B., Kraft, K. 2012. Assessment of genotoxicity of herbal medicinal products: A co-ordinated approach. *Phytomedicine* 19:472–476.
36. Kwan, Y., Shamarina, S., Sreenivasan, S., Kalsom U. 2017. Genotoxicity of selected Chinese medicinal plants, *Elephantopus scaber*, *Glycyrrhiza uralensis* and *Salvia miltiorrhiza* on *Allium cepa* assay. *Ann Pharmacol Pharm* 2:1070.
37. Moreira, D., Teixeira, S., Monteiro, M. 2014 Traditional use and safety of herbal medicines. *Rev Bras Farmacogn* 24:248–257.
38. Mosihuzzaman, M. 2012. Herbal medicine in healthcare—An overview. *Nat Product Commun* 7:807–812.
39. Muller, L., Kikuchi, Y., Probst, G., Schechtman, L., Shimada, H., Sofuni, T. 1999. ICH-Harmonised guidances on genotoxicity testing of pharmaceuticals: Evolution, reasoning and impact. *Mutat Res* 436:195–225.
40. Muller, L., Mauthe, R., Riley, C., et al. 2006. A rationale for determining, testing, and controlling specific impurities in pharmaceuticals that possess potential for genotoxicity. *Regul Toxicol Pharmacol* 44:198–211.
41. NAP. 2007. *Applications of Toxicogenomic Technologies to Predictive Toxicology and Risk Assessment*. National Academies Press, editor, Washington, DC.
42. OECD. 2014. OECD guidelines for testing of chemicals.
43. Ouedraogo, M., Baudoux, T., Stévigny, C., Nortier, J., Colet, J.-M., Efferth, T. 2012. Review of current and "omics" methods for assessing the toxicity (genotoxicity, teratogenicity and nephrotoxicity) of herbal medicines and mushrooms. *J Ethno Pharmacol* 140:492–512.

44. Posadzki, P., Watson, L., Ernst, E. 2013. Adverse effects of herbal medicines: An overview of systematic reviews. *Clin Med* 13:7–12.

45. Pelkonen, O., Xu, Q., Fan, T.P. 2014. Why is research on herbal medicinal products important and how can we improve its quality. *J Tradit Complement Med* 4:1–7.

46. Jordan, S., Cunningham, D., Marles, R. 2010. Assessment of herbal medicinal products: Challenges and opportunities to increase the knowledge base for safety assessment, *Toxicol Appl Pharmacol* 243:198–216.

47. Sponchiado, G., Adam, M.L. 2016. Quantitative genotoxicity assays for analysis of medicinal plants: A systematic review. *J Ethnopharmacol* 178:289–296.

48. Vaidya, A., Devasagayam T. 2007. Current status of herbal drugs in India: An overview. *J Clin Biochem Nutr* 41:1–11.

49. Valerio, L. 2009. *In silico* toxicology for the pharmaceutical sciences. *Toxicol Appl Pharmacol* 241:356–370.

50. Verma, S., Singh, S.P. 2008. Current and future status of herbal medicines. *Vet World* 1:347–350.

51. WHO. 2008. Herbal medicine research and global health: An ethical analysis. *Bull WHO* 86:577–656.

52. Zhou, J., Ouedraogo, M., Qu, F., Duez, P. 2013. Potential genotoxicity of traditional Chinese medicinal plants and phytochemicals: An overview. *Phytother Res* 27:1745–1755.

11 Deterrence of Genotoxicity

Brief Perspective

INTRODUCTION

The chemical processes uses reactive starting materials, intermediates and reagents, some of which are potential genotoxins. Therefore, APIs may restrain small levels of such probable GTIs, which has been increasingly demonstrated by both industry and regulatory bodies. The regulatory bodies have issued several guidelines in the last few years to support the need to manage the risk from potential GTIs, both during development and post-approval.

Genotoxic effects together with deletions, breaks or rearrangements can lead to cancer if the damage does not straight away lead to cellular death. Regions that are prone to breakage are called fragile sites which may result from genotoxic agents including pesticides. Some of the chemicals have the potential to set off fragile sites in chromosomal regions where oncogenes are present, which can lead to carcinogenic outcomes. In accordance with this finding, occupational publicity of various combinations of insecticides are positively correlated with amplified genotoxic damage in uncovered individuals [1]. DNA damage is not uniform in its severity across populations. Individuals diverge in their ability to set off or detoxify genotoxic substances, which leads to variability within the incidence of cancer among individuals. The distinction in skill to detoxify compounds is because of individuals' inherited polymorphism of genes in the metabolism of the chemical. Differences may also be ascribed to male or female discrepancy in the performance of DNA repair mechanisms [2].

The metabolism of a few chemical compounds results in the mechanised reactive oxygen species (ROS), which is a probable mechanism of genotoxicity. This is seen in the metabolism of arsenic, which produces hydroxyl radicals known to cause genotoxic effect [3]. Similarly, ROS have been involved in genotoxicity caused by particles and fibres. Genotoxicity of both non-fibrous and fibrous particles is characterised by excessive ROS production from inflammatory cells [4].

Flavonoids bear biochemical and pharmacological activities, which are likely detrimental and shielding. In addition, flavonoids can modulate xenobiotic metabolism. The concealed base for protection is penetrating with enzymes consisting of cytochrome p450, which performs a pivotal function in the metabolic activation of a wide variety of carcinogens [5].

Drugs currently being used as anti-mutagenic include busulfan, carmustine, etoposide, etc. Plant-derived polyphenolics and other chemical substances with antioxidant properties restrain the expression of genotoxic interest by using pro-oxidant chemical substances. *In-vitro* and *in-vivo* research with ionising radiation has shown that hydroquinone (HQ) has similar protective effects. The protecting impact of HQ is either because of enzyme induction or an immediate anti-oxidant effect of HQ against oxidants present in the diet [6].

Unlike ellagic acid (EA), ellagic acid peracetate (EAPA) has authenticated time-based reticence of liver microsomes catalysed AFB1 epoxidation, which is measured by binding of AFB1 to DNA. EAPA is tougher than EA in averting bone marrow and lung cells from AFB1 for genotoxicity.

EAPA operates in advance through microsomal acetoxy drug protein transacetylase (TAase) for intonation of the catalytic movement of explicit important proteins, such as CP450, NADPH, cytochrome C reductase and glutathione S transferase, possibly through protein acetylation [6].

HISTORY OF REGULATIONS UNTIL 2004

Regulatory authorities in the late 1990s and early 2000s recognised that genotoxic and carcinogenic impurities can pose a high risk. Therefore, they needed to be regulated mostly on a case to case basis. This field had not made enough advancement to allow for a unified regulatory guideline for which levels of GTIs might be acceptable or technically achievable. This was reflected in the ICH impurity guideline ICH Q3A "Impurities in New Drug Substances" issued in 1996. This guideline lacked a specific recommendation for mutagenic or carcinogenic impurities. However, it stated that, "discovery should be endeavoured for those prospective impurities which are accepted to be extraordinarily effective, producing noxious or pharmacologic consequences at a level lower than 0.1 percent".

In 2002, a revised ICH Q3A (R1) guideline "Impurities in New Drug Substances" was issued [7]. There was no specific address on how to handle mutagenic or carcinogenic impurities, it only suggested points to consider for potential GTIs, which stated that "lower thresholds may be appropriate for unusually toxic impurities". In 2004, the European Medicines Agency (EMEA) issued a draft titled "Guideline on the limits of Genotoxic Impurities" for comments [8]. It classified genotoxins into two categories:

- Compounds with substantiation for a threshold-related mechanism whereby the genotoxicity follows a threshold dose-response relationship.
- Compounds with no sufficient evidence for a threshold-related mechanism.

Guidelines introduced the concept of a threshold of toxicological concern (TTC) for such compounds. This threshold is based on the assumption of acceptance of 1 in 100,000 excess lifetime cancer risk and was based on the cancer risk of more potent known carcinogens [9].

Moreover, it is based on linear extrapolation from rodent carcinogenic doses and assumes a lifetime exposure. The TTC value established using this methodology was 1.5 mg/day for a given GTI. The guidance also generated a decision tree for the estimation and evaluation of the acceptability of GTIs to give the industry a risk assessment tool.

PHARMACEUTICAL INDUSTRY RESPONSE IN 2005 AND 2006

EMEA draft guideline issued in 2004 (vide supra) did not address the acceptable limits for GTIs during the clinical development phase prior to commercialisation. Thus, there was a concerted effort from the industries with various clinical trials by the FDA relating to GTIs in 2004. PhRMA working faction was ascertained which resulted in the periodical publication of a white paper which was the foundation for determining, testing and directing specific impurities in pharmaceuticals with a prospective for genotoxicity [10]. This PhRMA paper introduced two concepts:

The classification system for GTIs comprises five classes:

A. Genotoxic carcinogens,
B. Genotoxins with unknown carcinogenicity,
C. Impurities with a structural alert for genotoxicity,
D. Impurities with a structural alert similar to the parent API and
E. Impurities lacking an alerting structure.

Based on this, a strategy for impurity assessment based on the use of structure–activity relationships was proposed. Currently, this strategy has become the norm for most GTI-related risk assessments within the industry.

The concept of a staged TTC for clinical trial materials meant that the TTC was adjusted by taking into consideration the following:

• The dose – resulting in a lower TTC for higher doses – and
• The duration of the clinical trials – resulting in a higher TTC for a shorter duration.

Therefore, the staged TTC values can be appropriate for each individual compound where considerable GTIs are present. The exception would be highly potent carcinogens. Both concepts were introduced with the intention to provide some operational flexibility and a risk mitigation strategy during the development of a new drug.

In 2007, the EMEA issued their official Guideline on the Limits of Genotoxic Impurities [11]. It was followed in 2008 by a Question Answer Document [11].

The endorsement was made of a staged TTC concept from the PhRMA proposal (vide supra), but with reduced limits for each duration of exposure. In 2008, the FDA subjected an outline that served as a guide for industry pertaining to genotoxic and carcinogenic impurities in drug substances and products and certain recommended approaches [12]. It was inclusive of the staged TTC concept with the same limits for longer durations as the EMEA guidance. The FDA draft guidance was, by and large, quite well aligned with the EMEA guideline.

Some notable exceptions include:

• The FDA includes carcinogenic impurities (many carcinogens are non-genotoxic).
• The FDA includes additional safety margins for paediatric populations

CONTROL STRATEGIES

The control strategy has the following main points for consideration:

- Avoidance
- Adjusting API process
- Demonstrate GTI threshold mechanism above TTC level

AVOIDANCE

According to the EMEA guideline issued in 2006, the preferred control option was the avoidance of GTIs in APIs. For the processing chemist, the challenge with this approach was to be aware of the numerous side reactions that can produce reactive molecules at low levels that can further lead to potential GTIs. A good example of this is the widely used reaction of acids with alcohols. These reactions can produce assorted and copious alkylating agents, including nearly two dozen alkyl halides, esters of alkyl sulfonic acids (mesylates such as ethyl mesylate and methyl mesylate) and esters of aryl sulfonic acids (besylates and tosylates and esters of sulfuric acid), all of which have the potential to be GTIs [13]. In this case, it would seem noteworthy that evaluating and substituting such acids with alternative ones that do not produce GTIs and/or to limit the presence of the corresponding alcohols. Thus, it can be concluded that completely avoiding GTIs in efficient API processes is a challenge.

ADJUST API PROCESS

It is often unavoidable that the optimal chemical process to an API leads to GTIs. But some strategies can be used to adjust the process to optimise control over GTIs.

- The early placing of GTIs in the process.
    Early placing of GTIs will result in easy and more steps available for purging. In addition, the further away from the API forming step the GTIs can be effectively controlled, the more likely it will be that there is no need to include a specification for the GTIs.
- Strategic solid isolation placement in the process.
    Crystallisation is the most preferred way to purify intermediates and APIs. Therefore, API processes need to be adjusted such that crystallisations placed strategically in the process will support the potential to reduce and/or remove crucial GTIs.
    In addition to the above-discussed general strategies, it's worthwhile to think about applying detailed process understanding for the control of GTIs. A stimulating example of this was recently published [14]. The demonstration was done as to how to intricate process knowledge for multiple process parameters within the esterification reaction of alcohols with methanesulfonic acid that aids the understanding of the way to effectively control and limit genotoxic mesylate esters in API processes.

DEMONSTRATE GTI THRESHOLD MECHANISM ABOVE TTC LEVEL

An example has been proposed to demonstrate the GTI threshold mechanism. High levels of ethyl mesylate (EMS), a potential GTI, led to a recall of Roche's Viracept (nelfinavir mesylate) in 2007 [15]. The cause was eventually traced to a GMP failure in the manufacture of the API [15].

After cleaning a vessel with ethanol, the tank was not dried completely, leaving small amounts of ethanol in the tank. The tank was then charged with methane-sulfonic acid during the manufacture of the API. Over several months, significant amounts of EMS were formed that eventually contaminated the API.

While Roche immediately started investigating the origin of EMS and eventually solved the underlying manufacturing issue, they concurrently undertook an *in-vivo* rodent toxicology study with EMS. The surprising outcome was that a threshold of 2 mg EMS/kg for DNA damage was determined [15].

This threshold is approximately four orders of magnitude higher than the 1.5 mg/day TTC as per the EMEA guideline. Based on this finding, the EMEA accepted a higher TTC for EMS [15]. Therefore, it has been speculated that this result may ultimately provide a novel approach to guiding risk management for genotoxic impurities in pharmaceuticals [15]. This may be particularly true for monofunctional alkylating agents that react with DNA as soft nucleophiles via an SN2 mechanism in a similar fashion as EMS.

TWO SPECIAL CASES HAVE BEEN MENTIONED BELOW:

- SPECIAL CASE 1 – API STARTING MATERIALS

    Short API processes with few synthetic steps and/or only a few purification steps can pose a challenge if they contain GTIs. Regulators may pay special attention to these cases as there may be less opportunity for control over GTIs in such processes. This was highlighted during a recent conference [16]. As a consequence of having short API processes containing GTIs, the API starting material definition can become contentious. Moving back the GMP starting material in a process is one way that the regulators can try to assure additional process control over potential GTIs [16].

- SPECIAL CASE 2 – API DEGRADANTS

    Although the current regulations are silent on this issue, the topic has garnered attention in the industry and has recently received attention both in publications and in the more mainstream media [17]. In the absence of clear guidelines, a practical approach to evaluating GTIs in APIs was recently published [17].

It consists of the following steps:

a. List the functional groups present in the API.
b. List the most relevant degradation pathway for each functional group.

   c. Based on the results from A and B, predict the most likely degradant structures using chemical process knowledge and experience and evaluate the resulting structures for potential GTI alerts.

   d. If a GTI present, develop necessary control strategies.

The ongoing development of software tools for predicting degradant structures should benefit this approach. It can also be expected that the topic of GTIs in API degradants will gain more attention by regulators in the future.

PROFICIENT OPINION

In 2009, ICH issued an absolute model paper in which an M7 topic for control of GTIs was agreed upon [17]. This document listed the professed problems with interpreting the multiple current regulations that had evolved over the last decade as well as the main issues for which clarifications and guidance were desired. It was, therefore, proposed to move forward with work towards a new guideline, entitled "Assessment and Control of DNA Reactive (Mutagenic) Impurities in Pharmaceuticals to Limit Potential Carcinogenic Risk". This approach was authorised in 2010 and an expert working group (EWG) was established, which consisted of two members nominated by the six sponsors of the ICH, and one member nominated by Health Canada, WHO and EFTA as observers. This EWG in 2010 drafted a guideline for comments in late 2012 [17].

   The guidance addressed the following topics:

- Acceptable levels of GTIs during drug development
- Acceptable levels of GTIs for marketing
- Acceptable levels of GTIs with a strong likelihood to exhibit a threshold effect
- The merit of the TTC approach
- How to treat structurally related GTIs
- The data required to support higher daily intakes than those allowed by the TTC
- Genotoxic degradants and what control strategy to expect and apply

CONCLUSION

In the last 10 years GTIs have emerged as a noteworthy and innovative regulatory topic for the pharmaceutical industry. Therefore, it is expected that the planned issuance of the guidelines will further clarify and support an efficient regulatory path for GTIs during drug development and commercialisation of new APIs.

REFERENCES

1. Tice, R. 2000. Single cell gel / comet assay: Guidelines for in vitro and in vivo genetic toxicology testing. *Environ Mol Mutagen* 35:206–221.
2. Kihlman, B., Natarajan, A. 1984. Potentiation of chromosomal alterations by inhibitors of DNA repair. *Nucleic Acid Res* 13:319–339.

3. Luchnik, N., Fesenko, E., Orchinnikova, V. 1976. Critical periods of the mitotic cycle: Influence of aminopterin and thymidine on production of chromosomal aberrations by radiations in *Crepsis capillaris*. *Mutat Res* 34:367–388.

4. Bear, W., Teel, R. 2000. Effects of citrus flavonoids on mutagenicity of heterocyclic amines and on cytochrome P450 1A2 activity. *Anticancer Res* 5b:3609–3614.

5. Kumar, A., Tyagi, Y., Ponnan, P., et al. 2007. Ellagic acid peracetate is superior to ellagic acid in the prevention of genotoxicity due to aflatoxin B1 in bone marrow and lung cells. *J Pharm* 59:81–86.

6. Bhide, S., Shivapurkar, N., Gothoskar, S., et al. 1979. Carcinogenicity of betel quid ingredients feeding mice with aqueous extracts and the polyphenol fraction of betel nut. *Br J Cancer* 40:922.

7. http://www.ich.org/products/guidelines/quality/quality-single/article/impurities-in-new-drugsubstances.html (accessed March 20, 2020).

8. http://www.rsihata.com/updateguidance/emea/old/519902en.pdf (accessed March 20, 2020).

9. Kroes, R., Renwick, A.G., Cheeseman, M., et al. 2004. Structure-based thresholds of toxicological concern (TTC): Guidance for application to substances present at low levels in the diet. *Food Chem Toxicol* 42:65–83.

10. Müller, L., Mauthe, R., Riley, C.M., et al. 2006. A rationale for determining, testing, and controlling specific impurities in pharmaceuticals that possess potential for genotoxicity. *Regul Toxicol Pharmacol* 44:198–211.

11. http://www.ema.europa.eu/docs/en_GB/document_library/Scientific_guideline/2009/09/WC500002903.pdf (accessed March 20, 2020).

12. http://www.ema.europa.eu/docs/en_GB/document_library/Scientific_guideline/2009/09/WC500002907.pdf (accessed March 20, 2020).

13. http://www.fda.gov/downloads/Drugs/GuidanceComplianceRegulatoryInformation/Guidances/ucm079235.pdf (accessed March 20, 2020).

14. Sobol, Z., Engel, M.E., Rubitski, E., et al. 2007. Genotoxicity profiles of common alkyl halides and esters with alkylating activity. *Mutat Res* 633:80–94.

15. Teasdale, A., Elder, D., Chang, S., et al. 2013. Risk assessment of genotoxic impurities in new chemical entities: Strategies to demonstrate control. *Org Proc Res Dev* 17:221–230.

16. http://www.emea.europa.eu/humandocs/PDFs/EPAR/Viracept/38225608en.pdf (accessed on March 20, 2020).

17. Baertschi, S. 2011. Pharmaceutical stress testing: Predicting drug degradation. *Informa Healthc* 484–498.

12 Genotoxic and Non-Genotoxic Chemicals for Review of the Performance Improved Genotoxicity

INTRODUCTION

In 2007, [1] the recommendations of a workshop, organised and funded by the European Reference Laboratory for Alternatives to Animal Testing (EURL ECVAM) laid out the ways to reduce the frequency of "deceptive" or "extraneous" positive results (i.e. positive results that are found *in vitro* and are not prophetic and analytical of *in-vivo* genotoxic or carcinogenic action) meticulously in mammalian cell tests.

Several suggestions were identified for probable step-ups/variations to existing tests or novel tests that showed potential. Such enhancements or new assays need to show advanced specificity (i.e. give fewer "ambiguous" positive results) without compromising sensitivity (i.e. still detecting *in-vivo* genotoxins and DNA-reactive carcinogens) [2]. In recent years, several experimental initiatives have identified that the reliability of *in-vitro* mammalian cell tests (i.e. superior specificity while retaining high sensitivity) can be improved by using p53-competent human cells [3] and by choosing measures of cytotoxicity based on cell propagation [4] rather than simple methods, such as relative cell count or vital stains. However, if further improvement to *in-vitro* tests is to be evaluated, or if novel tests are to be developed, it is vital to have a reference set of chemicals to test where the outcomes (true positive, true negative or ambiguous positive) can be predicted based on results obtained under standard test conditions. EURL ECVAM, therefore, convened an expert working panel to identify the recommended lists of chemicals, which were then published [5].

The lists covered the following three sets of chemicals:

- **Group 1** are chemicals that are *in-vivo* genotoxins due to DNA-reactive and non-DNA-reactive mechanism; for example, the induction of aneuploidy and the restraining of topoisomerase. The majority of them are also recognised as carcinogens with a mutagenic approach of action, but a sub-class of apparent and probable aneugens has been established carcinogenic properties of which are not comprehensibly understood.

- **Group 2** are classified chemicals that should and regularly give negative results in *in-vitro* mammalian cell genotoxicity tests. Thus, chemicals in this group are, by and large, negative in *in-vivo* genotoxicity tests when tested and are non-DNA-reactive. Moreover, they are non-carcinogenic and rodent carcinogenic with an acknowledged non-mutagenic approach of action.
- **Group 3** are chemicals that give negative results in *in-vitro* mammalian cell genotoxicity tests, but they are also reported to bring about and provoke gene mutations in mouse lymphoma cells, chromosomal aberrations or micronuclei repeatedly at high concentrations and thus have high levels of cytotoxicity. Thus, chemicals in this group are, by and large, negative in *in-vivo* genotoxicity studies when they are tested, as well as negative in Ames test. They are both non-carcinogenic and rodent carcinogenic with a recognised non-mutagenic approach of action.

Since the publication of the EURL ECVAM, the recommended lists of genotoxic and non-genotoxic chemicals have become a key reference for test developers in the field of genotoxicity. The whole lists, or part of them, have been used in the evaluation of a large number of the newly developed genotoxicity assays [6–13].

The lists have also been used for the assortment of chemicals to determine whether alterations to existing protocols led to expansions of the assays in terms of the recital, and, in particular, regarding a superior specificity [14]. Among the novel and enhanced assays, many are high-throughput screening methods (HTS) [15–19]. In certain cases, existing tests have been miniaturised or automated, and the new accounts of these traditional tests need to be validated [20–23]. The optional chemical lists have likewise served to support the chemical selection of several international laboratories.

In this chapter, substances with a "mutagenic mode of action" refers to substances that are positive in the Ames test, whereas substances with a "non-mutagenic mode of action" are negative in the Ames test. Validation studies include the *in-vitro* and *in-vivo* test methods, among which the validation of the *in-vivo* Comet assay, the micronucleus and Comet assays in 3D reconstructed skin models, the Pig-a assay, the cell transformation assay in Bhas 42 and the micronucleus test in hen's eggs (HET-MN) have been performed [24–28].

Moreover, some projects that have used these lists have been carried out to investigate the ways to reduce misleading positive results in the *in-vitro* mammalian cell tests within the existing testing practices and to identify the appropriate parameters that need to be well-thought-out for the expansion of novel or enhanced tests [3,4,29]. The wide-ranging use of the chemicals and the references for test method design as sorted out by EURL ECVAM for the development and execution led to the revision and updating of the list at a workshop of EURL ECVAM at Ispra, Italy on 23rd October 2014.

Some questions pertaining to updating the chemical listings are as follows:

- The chemicals earlier recommended are still vindicated in light of new records or information previously not known.
- If the statistics did not support chemicals in the respective lists, should they be obliterated or apportioned to a different group?

- For chemicals that are obliterated, can their replacements be identified. Therefore, careful deliberation and contemplation need to be done as an indication that would sustain the inclusion of each chemical in each of the listings, with detailed justifications and supporting references, which is not an exhaustive list but considered as a representative.
- Absolute and reliable facts are still unavailable for all chemicals, although some have supplemented the Ames test data with novel tests in *Salmonella typhimurium* TA102. In these cases, with some gaps and inconsistencies, the assessment is based on an evidence-based approach. To develop evidence-based conclusions, data of various kinds are being considered more or less important.

The information to fabricate credible evidence that a chemical is a DNA-reactive carcinogen or it is an *in-vivo* genotoxin should be perceived with a novel or customised genotoxicity test, which may not necessarily be the same test required to decide whether a chemical is non-DNA-reactive and should not be detected.

The classification is based on *in-vivo* genotoxicity and DNA reactivity, while carcinogenicity acts as a supplementary criterion. The definition of DNA reactivity is primarily based on the results from bacterial mutagenicity tests, that is, Ames test. This indicates that a positive Ames test indicates DNA reactivity, whereas a negative Ames test is an indication of the absence of DNA reactivity. Conversely, aneugens that are non-DNA-reactive have now been included, but are genotoxic and, therefore, their inclusion is justified.

CONTEMPLATIONS FOR REVISING GROUP 1

Chemicals that are in-vivo genotoxins and DNA-reactive. These mutagenic carcinogens should be perceived as positive in in-vitro mammalian cell tests, that is, they are true positives.

According to the Carcinogenicity Potency Database [30], *p*-Chloroaniline, a Group I chemical, possessed non-carcinogenic property was classified as a Group 1 chemical previously [5]. Thus, it yielded mixed positive and negative results in *in-vivo* genotoxicity tests. On consideration of both the free base and the hydrochloride salt of the chemical, the behaviour must be the same in aqueous biological systems. Therefore, positive results in the *in-vivo* micronucleus test are attained, and it is, therefore, considered to be carcinogenic based on a more precise and rigorous National Toxicology Program (NTP) study.

Therefore, *p*-Chloroaniline, both free base and hydrochloride salt, has been retained in Group 1 and the supporting data have been restructured.

In contrast, a chemical which was also included as a Group 1 previously is chloramphenicol [5], which yields a negative result in the Ames test [31]. Neither TA102 nor *Escherichia coli* strain was employed for such tests. It was negative for micronucleus tests in several bone marrow studies in mice [32]. It is not considered to be a strong and sturdy candidate for inclusion in Group 1 and, therefore, chloramphenicol has been removed. The objective is not that all chemicals grouped in Group 1 should be tested to evaluate a novel or custom-made mammalian cell test system, but rather a selection of the chemicals belonging to diverse structural classes, and acting in an altered manner, should be available.

To increase the selections and alternatives, some chemicals were added to the list, such as 5-fluorouracil, cytosine arabinoside, which are nucleoside analogues, colchicine and vinblastine, which are aneugens. Mitomycin C and 4-nitroquinoline-N-oxide were also included in the list as they are proposed positive controls as per the OECD test guidelines for mammalian cell genotoxicity tests.

Group 1, therefore, now represents different classes and exhibits different modes of action. The attention is on chemicals that are considered to be DNA-reactive carcinogens and *in-vivo* genotoxins. Furthermore, *in-vivo* genotoxins, such as aneugens, and clastogens, such as topoisomerase inhibitors and nucleoside analogues, have been categorised that are not carcinogenic, and are negative or equivocal in the Ames test, but positive as detected for *in-vitro* mammalian cell tests. Therefore, the approach of action for tumour induction might not be the same as that leading to genotoxic responses.

CONTEMPLATIONS FOR REVISING GROUP 2

Non-DNA-reactive chemicals include non-genotoxic carcinogens that do give negative results in in-vitro mammalian cell genotoxicity tests and are considered as true negatives.

The group list has been revised after including novel or previously established data. Several chemicals have been, therefore, deleted from Group 2 [5] for the following reasons:

- *Phenanthrene* depicted reasonable evidence of being deficient of mutagenic/clastogenic effects in mammalian cells; therefore, phenanthrene may be considered as a bacterial mutagen. It yielded a positive response in TA100 at 12 μg/plate and above in the presence of high levels of Aroclor-induced (around 30%) rat liver S9 [33]. In addition, no *in-vivo* data in standard systems and no non-carcinogenic label is based on limited data; several studies have been described on mice using dermal, intraperitoneal or subcutaneous administration [34], which were all diminutive, and in many cases, group sizes were smaller, than normal. Therefore, it is possibly not a first-rate candidate for a true-negative group.
- *Fluometuron* is a herbicide; a review published by EFSA [35] on the soil metabolite of fluometuron, desmethyl fluometuron regarded them as clastogenic *in vitro* and found them to be inducing tumours in mice. Therefore, fluometuron may not be a first-rate contender for a true-negative group.
- *Cyclohexanone* is negative for standard Ames strains [36] but it is not tested in TA102 or *E. coli*. There are both negative and positive wrapping ups for induction of mutation in the mouse lymphoma tk mutation assay [37,38], as well as the induction of chromosomal aberrations (CA) for both *in vitro* [39–41] and *in vivo* [42]. At the same time, as most of the CA studies are old, the results may be subjected to questions. Owing to these uncertainties

with these statistics set, it was recommended and advocated that this is not a first-rate candidate for the class of true-negative group.

- *Progesterone* is found to be affirmative and positive for induction of micronucleus *in vitro* via an aneugenic mode of action [43], and therefore it is not appropriate for the class of the true-negative group.
- *Trisodium EDTA trihydrate* is negative for the Ames test strains [44,45] and in most *in-vitro* genotoxicity tests; however, it yielded a few positive results for EDTA and its salts. These results have been quite pragmatic, depending on assay type and cell type [46]. Moreover, in the bone marrow micronucleus assay acute doses of EDTA disodium salt, 5–20 mg/kg of body weight was used which induced a dose-dependent increase in the incidence of micronucleated polychromatic erythrocytes at 24 h sampling [47]. Therefore, it may not be a good contender for the class of the true-negative group.

Because some previous data did not take into account tests on *S. typhimurium* TA102 or *E. coli* WP2 strains, an extensive search was made for the additional data available for these strains. Such data were not available for chemicals which were to be booked and retained within Group 2. Therefore, in that case, some novel tests in TA102 were performed using standard plate incorporation methodology. Thus, only chemicals that are negative in Ames tests with an inclusion of the standard strains along with either TA102 or *E. coli* and thus have been included in Group 2.

Group 2, therefore, consists of non-DNA-reactive and Ames test-negative chemicals. The chemicals in sub-group I are inclusive of those chemicals which have depicted negative *in-vivo* genotoxicity data as well as are negative *in vitro* and non-carcinogenic. There are several non-carcinogens that are non-genotoxic *in vitro* as there is no published *in-vivo* genotoxicity data for such chemicals. Therefore, they have been included in sub-group II because the data suggested and recommended that they depicted negative result in any modified or novel *in-vitro* genotoxicity test systems. Sub-group III consists of non-genotoxic carcinogens for which the non-genotoxic status is sustained and maintained by negative *in-vitro* and *in-vivo* genotoxicity statistics.

CONTEMPLATIONS FOR REVISING GROUP 3

These are non-DNA-reactive chemicals that are inclusive of non-genotoxic carcinogens, metabolic poisons and others that give negative results in in-vitro mammalian cell genotoxicity tests, but they have been reported to bring about tk mutations in mouse lymphoma cells, CA or micronucleus, time and again either at high concentrations or at high levels of cytotoxicity, makes them misleading or irrelevant positives.

- At present, it is of utmost importance and significance to focus time and effort on Group 3 chemicals. In light of novel statistics, [3], several chemicals that were previously well thought out to be toxic either in p53-defective hamster cells or in p53-competent human cells, it was so only when the current protocols were undertaken.

- Some were non-toxic up to the maximum concentration that could be tested even though others induced toxicity but were still not genotoxic. Therefore, the previously described information yielded positive results in mammalian cells, which were rare events and were not easily reproduced; therefore, the circumstances under which the positive results were found have not been identified.
- Hence, these chemicals were deleted from Group 3 as the majority of mammalian cell tests with these chemicals yielded negative results. However, some chemicals have yielded positive results and therefore they have been published, as a result of which they were placed in Group 3.

Some chemicals that were in the past included in Group 3 [5] are positive, for example, in the Ames test. Therefore they must be considered to be potentially DNA-reactive or *in vivo*. There are several other reasons to query for the reported positive results in mammalian cells or the records due to the absence of unclear *in vivo* genotoxic or carcinogenic activity. These chemicals have been mentioned below:

- *Phthalic anhydride* in aqueous media is likely to form phthalic acid, which leads to varied pH effects and could confound the results [48].
- *Propyl gallate* was also reported positive in TA102 in the incidence of hamster S9 [49] and for induction of micronucleus and CA *in vivo* [50].
- *2,4-Dichlorophenol* was either equivocal or positive in the Ames test with hamster S9 [36,51] and was positive for CA in bone marrow and spermatocytes of mice [52], as well as was positive for Comet assay in stomach and colon of mice [53].
- *Sodium xylene sulfonate* was the only compound with a yield of negative carcinogenicity data as in case of rats but was not tested in mice, and there are no *in-vivo* genotoxicity data; therefore, the categorisation as a non-carcinogen may not be robust and vigorous, and there are no numbers and records to confirm whether it is non-genotoxic *in vivo*.
- *Curcumin* was reported to yield positive results not only in many mammalian cell tests, possibly as a result of apoptosis (cell death) [54], but also positive in p53-competent human cells [3]. It may be not quite useful to determine and ascertain whether novel or modified tests have shown some acceptable specificity. These chemicals have, therefore, been obliterated from the previous Group 3 list.

The following list of chemicals was considered for inclusion in Group 3, but were in the long run not incorporated for the following reasons:

- *Benzoate sodium* was the only compound with negative and downbeat carcinogenicity statistics in rats and has not been tested in mice, therefore, the categorisation as a non-carcinogen may not be robust and vigorous.
- *Benzoin* has yielded positive results for stimulation of UDS in rat hepatocytes [55] as well as for causing *hprt* mutations [56,57]. Such results are not typical for chemicals that may yield misleading positive results *in vitro*.

- *Methyl methacrylate* depicted very weakly positive *in-vivo* micronucleus results which were described in two separate parts of studies using intra-peritoneal dosing [51] and yielded clear positive micronucleus results after 1-day inhalation exposure [58]. Thus, it was categorised as a chemical negative for *in-vivo* genotoxicity.
- *Diphenhydramine HCl* although depicted negative substantiation of carcinogenic activity for either male or female mice, there was equivocal evidence of carcinogenic activity in males where a marginal increase in incidence of uncommon astrocytomas or gliomas and alveolar/bronchiolar neoplasm was observed, and in case of female rats there was a marginal increase in the prevalence of pituitary gland adenomas [51].

Note:

- In the deficiency and dearth of any *in-vivo* genotoxicity data, it is complicated to categorise any compound or substance as clearly non-carcinogenic and non-genotoxic *in vivo*. The prospect and chances of deletion of compounds need a methodical and comprehensive discussion about Group 3.
- There have been some reports of the induction of micronucleus *in vivo*, but there have been more robust studies which have shown negative results. Therefore, the weight of evidence has indicated a lack of accurate and factual genotoxic activity *in vivo*.

For these reasons, the ethyl acrylate was retained in Group 3.

To compensate for the deletions, several new chemicals have been added. In the revisions and additions to Group 3, it has been verified and checked for the negative Ames results. Earlier data were not inclusive of tests on either *S. typhimurium* TA102 or *E. coli* WP2 strains. Therefore, rummage has been around for additional statistics and information in these strains. As statistics were not available for some of the chemicals, these have been retained within or have been added to Group 3. Novel tests have been using standard plate incorporation methodology.

Thus, only those chemicals that have been revealed to be negative in Ames tests are included in the standard strains along with either TA102 or *E. coli*, and have been categorised in Group 3. Group 3, therefore, consists of non-DNA-reactive chemicals, which means that these chemicals will yield negative results for Ames test, which has formed the basis for their selection. These chemicals in this group have been chosen primarily because they give negative *in vivo* genotoxicity results.

It is optional and non-obligatory to include such chemicals into a validation study for a novel or modified *in-vitro* mammalian cell genotoxicity test to identify either negative results are obtained or to ascertain the circumstances for stipulation under which positive results are found. Hence, these chemicals have been arranged in order into three sub-groups.

- *Sub-group I* – it yields negative *in vivo* test results and therefore should be given precedence.

- *Sub-group II* – There is no adequate publication of *in-vivo* genotoxicity data about this group.
- *Sub-group III* – Only two chemicals, namely, sodium saccharin and ethylacrylate have been included in this group; they yielded mainly negative results for genotoxicity *in vivo* but with induction of tumours via a non-genotoxic method.

NOTE ON CONCORDANCE

It should be noted that "concordance" is a measure which is used for evaluating the performance of a test system. At present, it is the level of agreement between the results in the *in-vitro* tests relative to the expected *in-vivo* test outcomes for genotoxicity or carcinogenicity. Concordance relies profoundly on having reasonably equal numbers of carcinogens/genotoxins and non-carcinogens/non-genotoxins. Many prior concerted or validation trials have included numerous carcinogens but a very few non-carcinogens. It should be noted that while all of the chemicals mentioned in this chapter are commercially available, chemical laboratories should be aware of the quality and must obtain certificates of analysis for the test compounds regarding the details of purity and impurities. For good preparation of test chemical solutions, sufficient care should be taken in the suitable choice of the solvent/vehicle. Moreover, careful handling of chemicals is always imperative, especially when handling IARC group 1 chemicals or coded chemicals.

CONCLUSION

Subsequent to careful contemplation of the available literature, and, in particular, of data published since 2008, the lists of chemicals that can be used in the evaluation of modified or new mammalian cell genotoxicity assays have been updated. These lists arrange the chemicals according to the expected positive results *in vitro* or negative results, wherein the latter includes chemicals currently suspected of giving "deceptive" or "extraneous" positive results in existing assays. It has been noted that the chemicals that have been deleted since the publication are not necessarily considered to be "erroneous". However, they are considered to be less vigorous candidates for the groups that have been presented, and any data obtained should be viewed with some caution. Thus, it is widely anticipated that these lists may prolong and maintain to provide functional and efficient reference chemicals for scientists seeking to modify existing assays or introduce new ones.

REFERENCES

1. Kirkland, D., Pfuhler, S., Tweats, D., et al. 2007. How to reduce false positive results when undertaking in vitro genotoxicity testing and thus avoid unnecessary follow up animal tests: Report of an ECVAM workshop. *Mutat Res* 628:31–55.
2. Kirkland, D., Aardema, M., Henderson, L., et al. 2005. Evaluation of the ability of a battery of 3 *in vitro* genotoxicity tests to discriminate rodent carcinogens and non carcinogens. I. Sensitivity, specificity and relative predictivity. *Mutat Res* 584:1–256.

3. Fowler, P., Smith, R. Young, J., et al. 2012. Reduction of misleading (false) positive results in mammalian cell genotoxicity assays. I. Choice of cell type. *Mutat Res* 742:11–25.

4. Fowler, P., Smith, R., Young, J., et al. 2012. Reduction of misleading (false) positive results in mammalian cell genotoxicity assays. II. Importance of accurate toxicity measurement. *Mutat Res* 747:104–117.

5. Kirkland, D., Kasper, Müller, P. 2008. Recommended lists of genotoxic and non genotoxic chemicals for assessment of the performance of new or improved genotoxicity tests: a follow-up to an ECVAM workshop. *Mutat Res* 653:99–108.

6. Birrell, L., Cahill, P., Hughes, C. 2010. GADD45a-GFP GreenScreen HC assay results for the ECVAM recommended lists of genotoxic and non genotoxic chemicals for assessment of new genotoxicity tests. *Mutat Res* 695:87–95.

7. Mizota, T., Ohno, K., Yamada, T. 2011. Validation of a genotoxicity test based on p53R2 gene expression in human lymphoblastoid cells. *Mutat Res* 724:76–85.

8. Berthelot-Ricou, A., Perrin, J., Di Giorgio, C. 2011. Comet assay on mouse oocytes: An improved technique to evaluate genotoxic risk on female germ cells. *Fertil Steril* 95:1452–1457.

9. Hendriks, G., Atallah, M., Morolli, B. 2012. The ToxTracker assay: Novel GFP reporter systems that provide mechanistic insight into the genotoxic properties of chemicals. *Toxicol Sci* 125:285–298.

10. Zwart, E., Schaap, M., van den Dungen, M. 2012. Proliferating primary hepatocytes from the pUR288 lacZ plasmid mouse are valuable tools for genotoxicity assessment in vitro. *Environ Mol Mutagen* 53:376–383.

11. Rajakrishna, L., Krishnan Unni, S., Subbiah, M. 2014. Validation of a human cell based high throughput genotoxicity assay 'Anthem's genotoxicity screen using ECVAM recommended lists of genotoxic and non genotoxic chemicals. *Toxicol Vitr* 28:46–53.

12. Ireno, I., Baumann, C., Stöber, R., et al. 2014. Fluorescence based recombination assay for sensitive and specific detection of genotoxic carcinogens in human cells. *Arch Toxicol* 88:1141–1159.

13. Bryce, S., Bemis, J., Mereness, J. 2014. Interpreting in vitro micronucleus positive results: simple biomarker matrix discriminates clastogens, aneugens, and misleading positive agents. *Environ Mol Mutagen* 55:542–555.

14. Le Hegarat, L., Dumont, J., Josse, R. 2010. Assessment of the genotoxic potential of indirect chemical mutagens in HepaRG cells by the comet and the cytokinesis block micronucleus assays. *Mutagenesis* 25:555–560.

15. Smart, D., Ahmedi, K., Harvey, J. 2011. Genotoxicity screening via the H2AX by flow assay. *Mutat Res* 715:25–31.

16. Westerink, W., Stevenson, J., Horbach, G. 2010. The development of RAD51C, Cystatin A, p53 and Nrf2 luciferase reporter assays in metabolically competent HepG2 cells for the assessment of mechanism based genotoxicity and of oxidative stress in the early research phase of drug development. *Mutat Res* 696:21–40.

17. Khoury, L., Zalko, D., Audebert, M. 2013. Validation of high throughput genotoxicity assay screening using H2AX in cell western assay on HepG2 cells. *Environ Mol Mutagen* 54:737–746.

18. Garcia-Canton, C., Anadon, A., Meredith, C. 2013. Assessment of the in vitro H2AX assay by High Content Screening as a novel genotoxicity test. *Mutat Res* 757:158–166.

19. Van der Linden, S., Von Bergh, A., Van Vught Lussenburg, B., et al. 2014. Development of a panel of high throughput reporter gene assays to detect genotoxicity and oxidative stress. *Mutat Res* 760:23–32.

20. Lukamowicz Rajska, M., Kirsch Volders, M., Suter, W., et al. 2012. Miniaturized flow cytometry based *in vitro* primary human lymphocyte micronucleus assay validation study. *Environ Mol Mutagen* 53:260–270.

21. Thougaard, A., Christiansen, J., Mow, T., et al. 2014. Validation of a high throughput flow cytometric in vitro micronucleus assay including assessment of metabolic activation in TK6 cells. *Environ Mol Mutagen* 55:704–718.

22. Tilmant, K., Gerets, H., De Ron, P., et al. 2013. The automated micronucleus assay for early assessment of genotoxicity in drug discovery. *Mutat Res* 751:1–11.

23. Westerink, W., Stevenson, J., Lauwers, A., et al. 2009. Evaluation of the Vitotox and Radar Screen assays for the rapid assessment of genotoxicity in the early research phase of drug development. *Mutat Res* 676:113–130.

24. Uno, Y., Kojima, H., Omori, T., et al. 2015. JaCVAM organized international validation study of the in vivo rodent alkaline comet assay for detection of genotoxic carcinogens: II. Summary of definitive validation study results. *Mutat Res* 786:45–76.

25. Reus, A., Reisinger, K., Downs, T., et al. 2013. Comet assay in reconstructed 3D human epidermal skin models investigation of intra and inter laboratory reproducibility with coded chemicals. *Mutagenesis* 28:709–720.

26. Dertinger, S.D., Phonethepswath, S., Avlasevich, S.L., et al. 2012. Efficient monitoring of in vivo pig-a gene mutation and chromosomal damage: Summary of 7 published studies and results from 11 new reference compounds. *Toxicol Sci* 130:328–348.

27. Sakai, A., Sasaki, K., Hayashi, K., et al. 2011. An international validation study of a Bhas 42 cell transformation assay for the prediction of chemical carcinogenicity. *Mutat Res* 725:57–77.

28. Greywe, D., Kreutz, J., Banduhn, N., et al. 2012. Applicability and robustness of the hen's egg test for analysis of micronucleus induction (HET-MN): Results from an interlaboratory trial. *Mutat Res* 747:118–134.

29. Fowler, P., Smith, R., Smith, K., et al. 2014. Reduction of misleading (false) positive results in mammalian cell genotoxicity assays. III: Sensitivity of human cell types to known genotoxic agents. *Mutat Res* 767:28–36.

30. Gold, L. The carcinogenic potency project. http://toxnet.nlm.nih.gov/cpdb/, 2004 (accessed March 20, 2020).

31. Mortelmans, K., Haworth, S., Lawlor, T., et al. 1986. Salmonella mutagenicity tests. II. Results from the testing of 270 chemicals. *Environ Mutagen* 8:1–119.

32. Morita, N., Asano, T., Awogi, Y., et al. 1981. Evaluation of the rodent micronucleus assay in the screening of IARC carcinogens (groups 1, 2A and 2B). The summary report of the 6th collaborative study by CSGMT/JEMS.MMS. *Mutat Res* 389:3–122.

33. Oesch, F., Bücker, M., Glatt, H. Activation of phenanthrene to mutagenic metabolites and evidence for at least two different activation pathways. *Mutat Res* 81:1–10.

34. IARC monographs on the evaluation of the carcinogenic risk of chemicals to humans Polynuclear aromatic compounds, Part 1, Chemical, environmental and experimental data, International Agency for Research on Cancer, Lyon, France. 1983.

35. European Food Safety Authority (EFSA). 2011. Conclusion on the peer review of the pesticide risk assessment of the active substance fluometuron, *EFSA J* 9:54.

36. Zeiger, E. 1997. Genotoxicity database, In: Gold, L.S., Zeiger, E (eds) *Handbook of Carcinogenic Potency and Genotoxicity Databases*. CRC Press Inc., Boca Raton. 687–729.

37. Mitchell, A., Auletta, A., Clive, D., et al. 1997. The L5187/tk+/− mouse lymphoma specific gene and chromosomal mutation assay. A phase III report of the U.S. Environmental Protection Agency Gene Tox Program. *Mutat Res* 394:177–303.

38. Seifried, H., Seifried, R., Clarke, J., et al. 2006. A compilation of two decades of mutagenicity test results with the Ames *Salmonella typhimurium* and L5178Y mouse lymphoma cell mutation assays. *Chem Res Toxicol* 19:627–644.

39. Collin, J. 1971. Cytogenetic effect of sodium cyclamate, cyclohexanone and cyclohexanol. *Diabete (French)* 19:215–221.

40. Lederer, J., Collin, J., Pottier Arnould, A. 1971. Cytogenetic and teratogenic action of cyclamate and its metabolites. *Therapeutique* 47:357–363.
41. Dyshlovoĭ, V., Boĭko, N., Shemetun, A. 1981. Cytogenetic action of cyclohexanone. *Gig Sanit (Russian)* 5:76–77.
42. De Hondt, H., Temtamy, S., Abd Aziz, K. 1983. Chromosomal studies on laboratory rats (*Rattus norvegicus*) exposed to an organic solvent (cyclohexanone). *Egypt J Genet Cytol* 12:31–40.
43. Kayani, M., Parry, J. 2008. The detection and assessment of the aneugenic potential of selected oestrogens, progestins and androgens using the in vitro cytokinesis blocked micronucleus assay. *Mutat Res* 651:40–45.
44. Zeiger, E., Anderson, B., Haworth, S., et al. 1988. Salmonella mutagenicity tests. 4. Results from the testing of 300 chemicals. *Environ Mol Mutagen* 11:1–158.
45. Dunkel, V., Zeiger, E., Brusick, D., et al. 1984. Reproducibility of microbial mutagenicity assays: VI. Tests with *Salmonella typhimurium* and *Escherichia coli* using a standardized protocol. *Environ Mutagen* 6:1–251.
46. Lanigan, R., Yamarik, T. 2002. Final report on the safety assessment of EDTA, calcium disodium EDTA, diammonium EDTA, dipotassium EDTA, disodium EDTA, TEA EDTA, tetrasodium EDTA, tripotassium EDTA, trisodium EDTA, HEDTA, and trisodium HEDTA - cosmetic ingredient review expert panel. *Int J Toxicol* 21:95–142.
47. Narasimhamurthy, K. 1991. Assessment of *in vivo* mutagenic potency of ethylene diamine tetracetic acid in albino mice. *Food Chem Toxicol* 29:845–849.
48. SIDS Initial Assessment Report for SIAM 20 on Phthalic anhydride, UNEP Publications. http://www.chem.unep.ch/irptc/sids/OECDSIDS/85449.pdf, 2005 (accessed March 20, 2020).
49. Fujita, H., Nakano, M., Sasaki, M. 1990. Mutagenicity test of food additives with *Salmonella typhimurium* TA97 and TA102. III. *Chemsitry* 39:343–350.
50. Shelby, M., Witt, K. 1995. Comparison of results from mouse bone marrow chromosome aberration and micronucleus tests. *Environ Mol Mutagen* 25:302–313.
51. NTP. NTP website at http://tools.niehs.nih.gov/cebs3/ui (accessed March 20, 2020).
52. Amer, S.M., Aly, F.A. 2001. Genotoxic effect of 2,4-dichlorophenoxy acetic acid and its metabolite 2,4-dichlorophenol in mouse. Mutat Res 494:1–12.
53. Sasaki, Y.F., Sekihashi, K., Izumiyama, F. et al. 2000. The comet assay with multiple mouse organs: comparison of comet assay results and carcinogenicity with 208 chemicals selected from the IARC monographs and U.S. NTP carcinogenicity database. Crit Rev Toxicol 30:629–799.
54. Mientières, S., Biola, A., Pallardy, M. et al. 2003. Using CTLL-2 and CTLL-2 bcl2 cells to avoid interference by apoptosis in the in vitro micronucleus test. Environ Mol Mutagen 41:14–27.
55. Glauert, H.P., Kennan, W.S., Sattler, G.L. et al. 1985. Assays to measure the induction of unscheduled DNA synthesis in cultured hepatocytes. Prog Mutat Res 5:371–373.
56. Lee, C.G., Webber, T.D. 1985. The induction of gene mutations in the mouse lymphoma L5178Y TK+/− assay and the Chinese hamster V79HGPRT assay. Prog Mutat Res 5:5457–5554.
57. Kuroda, Y., Yokoiyama, A., Kada, T. 1985. Assays for the induction of mutations to 6-thioguanine resistance in Chinese hamster V79 cells in culture. Prog Mutat Res 5:537–542.
58. De Araújo, A.M. Alves, G.R. Avanc, G.T. et al. 2013. Assessment of methyl methacrylate genotoxicity by the micronucleus test. *Braz. Oral* Res 27:31–36.

13 Genotoxic Impurities in Pharmaceuticals

INTRODUCTION

In pharmaceutical products, impurities are defined as those compounds/substances that endow non-affirmative therapeutic benefit but do have some potential owing to which they can cause mild-to-severe adverse effects. Therefore, for such instances, these impurity levels need to be controlled and managed such that these pharmaceutical products are satisfactorily and sufficiently safe and sound upon their administration to humans [1].

Impurities have a greater say or impact on the safety and time taken for development for the production of medicinal products, as well as with the consideration of the sales and marketing of drugs. For example, the period required for the development of drugs can be significantly and considerably increased when it is obligatory to characterise and remove impurities to acceptable levels. In some illustrations, marketing of these drugs has been impacted upon their removal from the market, which is primarily attributed to contamination by a genotoxic substance [1].

As per the International Conference on Harmonization (ICH) guidelines, impurities that are related to drug substances can be broadly classified into three main categories:

- Organic impurities
- Inorganic or elemental impurities
- Residual solvents

Within these groupings, genotoxic impurities (GIs) form an exceptional case that presents a significant and noteworthy safety risk, even at low concentrations, possibly because they may be mutagenic and are therefore detrimental to the DNA. Therefore, they can either direct mutations or form the grounds for cancer.

There have been many discussions regarding the definition of genotoxins and genotoxicity. For this purpose, the primary reference is to the *"ICH S2 (R1) Guideline 1"*. Thus, genotoxicity or genetic toxicity is understood as *"a broad expression that refers to any deleterious, detrimental alteration in the genetic material apart from of the approach by which the alteration is induced"*.

GIs can also be expressed as "an *impurity that has been inveterate and ascertained to be genotoxic in an appropriate, apposite and suitable genotoxicity test models, e.g., bacterial gene mutation (Ames) test"*.

A potential genotoxic impurity (PGI) can be described as an *"Impurity that gives an idea about the structural alert(s) for genotoxicity but it has not been tested in any experimental test model"*. The potential relates to the genotoxicity of the compounds/substances, and not to the incidence or dearth of the impurity [1].

DOGMATIC IMPLICATIONS: REGULATIONS AND GUIDELINES

The impurities present in drug substances and drug products are prospectively toxic and have no assistance for the patients [1]. Following are the diverse sources of impurities present in drug substances:

- Preparatory materials and their impurities
- Reagents and catalysts employed
- Solvents
- Intermediates formed
- Excipients and their impurities
- Leachables
- Degradation artefacts

To save from harm and shield patients, the impurities present in drugs must be reduced to the standard safety periphery. There are numerous dogmatic guiding principles, strategies and position papers with the primary attention on domineering the amount of impurities present in medicinal products within specified limits [2].

The impurity guidelines have principally been developed by the ICH. For example:

- ICH Q3A15 guideline regulates for the impurities present in new drug substances, with brinks for accounting, categorising and becoming certified as impurities.
- ICH Q3B16 is a similar and comparable guideline regarding impurities in new drugs.
- ICH Q3C17 is a guideline that regulates and controls for the residual solvents employed, and is the foremost instance that the ICH has applied substance-specific limits.
- ICH Q3D guideline deals with the elements and limits for heavy metal impurities.

Depending on the impending and prospective risk associated with human health, residual solvents employed have been categorised into the following three classes:

- Class I solvents, they should be avoided.
- Class II solvents, they have legitimate daily exposure limits.
- Class III solvents, they have no health-based revelation limits, if daily revelation is ≤50 mg/day.

The contemporary approach is to define element-specific limits amounting in finished drugs and for officially recognised daily revelation limits [2].

One issue associated with GIs, in particular, is that the synthesis of drug substances repeatedly necessitates the exploitation of reactive materials which have the capability and competence to interact with human DNA to cause mutation and cancer, even at particularly lowest levels [2].

Therefore, GIs should be circumvented, and if circumvention is not possible, then their levels need to be reduced below a distinct threshold. Different groups from industries and regulatory authorities have expanded guidelines purposely and distinctively while addressing GIs [2].

Interrelated and interconnected guidelines have been discussed in the subsequent sections of this chapter and summarized in Table 13.1. Thus, this chapter presents an overview of imperative guidelines as well.

GUIDELINES FOR GENOTOXIC IMPURITIES, CONTROL, TESTING, AND RISK ASSESSMENT [3]

PHRMA POSITION PAPER

The pharmaceutical industry's identification of the limits for GIs and the prerequisite obligation of official guidelines were first combined with FDA inspection followed by enforcement of the practices regarding genotoxins. This prompted the industries to retort with a resolute response [4].

In 2004, PhRMA formed a task force to discuss GIs and its effort resulted in the publication of a spot paper in regulatory toxicology and pharmacology as an underlying principle for determination, test and control of unambiguous impurities in pharmaceuticals that have power over potential for causing genotoxicity in 2006 [4].

This paper was developed along with the European Medicines Agency's (EMA) inventiveness for development of related authorised guidelines, and as a result of which there is an overlay in the approach. For instance, the PhRMA document adopted the conception of a threshold of toxicological concern (TTC) from the draft of EMA. The ultimate EMA guideline adopted commendations and counsel for the staged TTC approach from the PhRMA paper [5].

One major disparity and distinction between them is their prominence on the implementation, with the official regulatory guideline. The EMA document has lesser details for the operation than the PhRMA paper. In short, it can be summarised that both documents are harmonising and corresponding in their approaches and importance [5].

EUROPEAN MEDICINES AGENCY (EMA) GUIDELINES

The EMA is the groundbreaking dogmatic authority that has given detailed guidelines regarding how GIs need to be handled. The gigantic task of development of draft was assigned to the Safety Working Party (SWP) of the Committee for Proprietary Medicinal Products (CPMP), which is now the Committee for Medicinal Products for Human Use (CMPH). The SWP committee held that impurities with genotoxic potential are considered as a special case and they are not covered under the Q3A and Q3B guidelines [6].

TABLE 13.1
Guidelines for the Control of Genotoxic Impurities

S. No.	Topics	Title
	Guidelines for the control of Genotoxic Impurities	PhRMA position paper is entitled with a *raison d'être* for the determination, testing and control of some specific impurities in pharmaceuticals that hold potential for genotoxicity.
		It introduced the vital and significant concept of the five class classification of the impurities and for the staged impurity threshold for short-term exposure.
		EMA drafted a document in 2006, with several consultations. It was accompanied by the related questions and answers along with the document. It is, therefore, considered as the most complete and comprehensive regulatory document which introduced the concept and theoretical model for the TTC.
		EMA in collaboration with SWP prepared the questions and answers on the guideline regarding the limits of GIs.
		FDA prepared the draft of the recommended approaches as guidance for the industry which takes into account the genotoxic and carcinogenic impurities in drug compounds/substances and drug products: In general, it is in the line up with that of the EMA guideline.
		ICH M7 guideline is related to the appraisal, evaluation and control of mutagenic DNA-reactive impurities in pharmaceuticals to the potential limit of the carcinogenic risk.
	Guidelines for genotoxicity testing	ICH S2 guideline is related to the genotoxicity testing and subsequent interpretation of data for those pharmaceuticals that are intended for human use. This document is a combination of previous guidelines of ICH: ICH S2A and S2B. Thus, it is regarded as the global document for genotoxicity testing.
		EMA issued a guideline for the assessment of genotoxicity of herbal substances/preparations. This guideline illustrates and expresses, in general, a framework and enlists the practical approaches for testing the probable genotoxicity of herbal substances/preparations, as well as the interpretation of results.
	Risk assessment for genotoxic and carcinogenic substances	The European Commission Health and Consumer Protection Directorate laid the document for the methodology involved in general risk assessment and subsequent approaches for genotoxic and carcinogenic substances.

Therefore, the intention addressed the loopholes and gaps in the ICH guidelines regarding the managing of impurities, and predominantly for those impending impurities which are likely to be extraordinarily potent, such as the PGIs.

The committee further advised and recommended the addition of risk based on the safety factors up to the limits which are similar as per the ICH and the approach practised in Q3C for residual solvents. The first draft, titled as the position paper on the limits of GIs, was out for discussion in 2002, which was entirely discharged in 2004 [7].

Since this publication, the authorised and certified acronym for the European Medicines Agency was modified from EMEA to EMA. Therefore, for uniformity, the use of acronym EMA has been followed hereafter [7].

In 2006, the EMA's CHMP distributed and circulated the final guideline on the limits of GIs. The CHMP SWP to a large extent augmented the guideline by issuing and publishing several question and answers for further elucidation.

The document needs be appropriate to GIs in new drug compounds/substances. It is also appropriate to new applications for subsisting active substances, where consideration of the course of synthesis, process control and impurity silhouette does not provide level-headed declaration that no novel or elevated and advanced levels of GIs are pioneered as weighing against the products contemporarily authoritative when containing the same active substance [8].

In the current circumstances and framework, the categorisation of a compound as genotoxic in broad-spectrum ways that there are optimistic pronouncements ascertained using *in-vitro* or *in-vivo* genotoxicity tests with the main attention on DNA-reactive substances that have a prospective for direct DNA damage. Secluded *in-vitro* pronouncements may be calculated for *in-vivo* importance in follow-up testing [8].

FOOD DRUG ADMINISTRATION (FDA)

US FDA published draft pertaining as a guidance for industry, which was related to the genotoxic and carcinogenic impurities in drug compounds/substances and drug products. It was issued in December 2008. This FDA guidance gives precise and explicit commendations regarding the safety prerequisite of the impurities with either a known or a suspected genotoxic or carcinogenic probability. The guidance portrays and explains a diverse variety of approaches both for the characterisation and reduction of the potential risk of cancer with the patient exposure to genotoxins and carcinogenic impurities. The advances were comparable and related to the EMA guidelines as mentioned in Table 13.2 [8].

INTERNATIONAL COUNCIL FOR HARMONIZATION (ICH)

The foremost guideline regarding genotoxicity was initiated by the ICH in July 1995 designated as S2A. It relates the guidelines on the unambiguous facets of regulatory genotoxicity tests for pharmaceuticals. At that time, the issued guideline provided specific guidance and suggestions for *in-vitro* and *in-vivo* tests on the assessment and estimation of test results.

TABLE 13.2
FDA Guidance Measures and Stages for Minimisation of Risk

Step	Action
1	Alteration to either the synthesis or purification procedure to minimise the creation pattern and maximise the amputation and exclusion of the significant impurity.
2	Allowance of a maximum daily revelation for the target of 1.5 µg per day of the appropriate and germane impurity as a wide-ranging objective.
3	Further characterisation of the genotoxic and carcinogenic menace to better sustain and maintain the apposite impurity stipulation, either for superior or subordinate values.

The second guideline was issued in 1997 which was designated as S2B. It was designated for genotoxicity testing and sets of standard battery for testing for pharmaceuticals.

In 2013, the M7 guideline for the analysis of structure–activity relationships (SAR) studies for genotoxicity was published. After this, M7 (R1) guideline was discharged in two steps in June 2015 (as step 2) and in May 2017 (as step 4). The M7 (R2) guideline is up for revisions. This guideline integrates and slots in as acceptable limits, acceptable intakes, permitted daily exposures (PDEs) for novel DNA-reactive mutagenic impurities. Where the acceptable limits were revised for impurities which are already listed in the addendum. Therefore, the resulting new data on availability resulted in ICH M7 (R2) version guideline [9,10].

CLASSES OF GENOTOXIC IMPURITIES

GIs are classified based on their risk estimation which involves a preliminary analysis of definite and prospective impurities by searching the database and literature investigation for carcinogenicity and bacterial mutagenicity data after which they are divided into diverse classes from 1, 2 or 5.

If the figures, data or information for such a classification is not handy, then an estimation and evaluation with the help of SAR studies will focus on bacterial mutagenicity calculations, which can then be executed. Therefore, they are classified into either Class 3, 4 or 5.

Each class is defined and characterised as below [10].

Class 1 – These impurities have well-established mutagenic and carcinogenic data and are recognised and identified to carry serious risk and need to be eliminated by some modification in the process. If this is not probable or doable, these impurities are to be edged at TTC as the ultimate decision.

Class 2 – Though these impurities have well-established mutagenic data, their prospective potential as carcinogens is not known. Hence, these impurities need to be managed using the TTC approach.

Class 3 – These impurities have a structural alert which is unrelated to the structure of the drug compounds/substances with some unknown genotoxic potential. Based on the functional groups present within the molecule, they can be classified as genotoxic. The toxicity of these impurities is based on SAR studies.

Class 4 – These impurities are structurally similar to the structure of drug compounds/substances and contain the functional group or moiety that potentially shares the structure with the parent and is considered to be non-genotoxic.

Class 5 – These impurities have no structural alert and the substantiation point towards the non-existence of genotoxicity. These compounds are, therefore, treated as customary impurities and are organised and managed as per the ICH guidelines.

THRESHOLD OF TOXICOLOGICAL CONCERN (TTC) APPROACH

In 2006, EMEA proposed that the TTC, as a toxicological appraisal and evaluation, of the entire impurities is very complicated, therefore the brink or TTC was employed to control the GIs. The acceptable and tolerable intake of a mutagenic impurity of 1.5 µg per person per day is well-thought-out to be related with an insignificant risk, with a theoretical surplus cancer risk of <1 in 100,000 over a lifetime of exposure. This advance would usually be utilised for the mutagenic impurities at hand in pharmaceuticals for long-term treatment of more than 10 years and where negative carcinogenicity data are available, such as Classes 2 and 3. The frontier for individual genotoxic impurity can be computed by the formula described below [11].

Limit(in ppm) = 1.5 (microgram per day)/maximum daily dose (gram per day)

The consideration that all drugs are not worn out for long-term treatment, the novel concept of staged TTC was introduced by Mueller. This concept takes into account the fact that the extent of exposure is a vital and important factor, which has a say and impact on the prospective of a carcinogenic response.

RISK ASSESSMENT OF IMPURITY

The management and control of GIs is an exceptionally significant activity for any drug compound/substance or drug product. A vital element of this is risk evaluation and appraisal. The primary constituents of such estimation are to spotlight explicitly on the effectual use of *in-silico* assessment tools. For this impurity, the first step is the identification of structure, and the second step is the conduction of GIs assessment. If the defined impurity structure shows no sign of an alert, an impurity is said to be controlled and designated as a normal impurity as per the ICH Q3 guidelines. But if the impurity shows some structure alert, then it is focused on the Ames test. Hence, there can be two cases as discussed below:

- A negative result in this case for Ames test pilots to manage, organise and control the impurity as per ICH Q3 guideline.
- If Ames test is found to be positive, then the impurity is restricted to the safety level as apiece the TTC approach.

Information regarding the chemical structure of the impurity and its approach towards the formation mechanism is essential and imperative to gauge its toxicological connotation and repercussions, thus it will result in the improvement of the synthetic chemical processes to reduce or remove the impurity [11].

SECLUSION OF GENOTOXIC IMPURITIES

For the seclusion of the impurities, predominantly, chromatographic techniques are employed along with some classical techniques. Seclusion of impurities is very necessary to establish its structure and toxicity. The following methods have been worn out for seclusion of such GIs [12].

- Solid phase extraction
- Column chromatography
- Flash chromatography

- Supercritical fluid chromatography
- Thin-layer chromatography
- Capillary electrophoresis
- Preparative high-pressure liquid chromatography
- Accelerated solvent extraction
- Liquid-liquid extraction

The seclusion of such impurities should be done in the beginning based either on simple extraction which is uncomplicated or partition methods. Another probable criteria or plan to extract impurities is selective. This selectivity is based on acidity, basicity or neutrality.

The extraction processes, by and large, engross and rivet the liquid-liquid extraction, where there are two phases, one is an aqueous solution and the other is a non-polar organic phase. In the chromatographic method, the required impurity peak or band is separated, concentrated and isolated. This can be completed by using a simple chromatographic column which is prepared in the laboratory or using an instrument, such as the flash chromatograph and a preparative chromatograph [12].

ANALYSIS OF IMPURITIES

Identification, discovery and recognition of impurity are one of the key and vital activities of impurity profiling procedure. Thus, the objective is to categorise and spot the chemical structures of impurities which are present in the drug substances or are being observed in the stability studies exceeding a particular threshold.

This process of identification of such pharmaceutical impurities can be performed by application of various spectroscopic techniques, such as ultraviolet (UV), infrared, mass spectrometry and nuclear magnetic resonance. The quantitation can be performed with the application of the following chromatographic techniques:

- High-performance liquid chromatography endowed with a mass detector or a UV detector
- Gas chromatography equipped with a mass detector, flame ionisation detector or electron capture detector
- Supercritical fluid chromatography
- Thin-layer chromatography
- High-performance thin-layer chromatography

The ICH guidelines show that all the GIs present in drug compound/substance need to be identified if present at or above a certain or specified limit, which is referred to as identification threshold [13].

ANALYTICAL METHOD DEVELOPMENT

The fundamental assignment for the development of an analytical method for GIs is to develop a method which can perceive, discover and identify the GIs at tracing levels and find below the TTC. The developed analytical method should have lesser unpredictability and inconsistencies by conducting a sequence of guarded experimentation to compose and formulate the quality and safety of drug products.

As global regulatory requirements have become sterner and stringent, analytical methods for global products must be able to convene and congregate up to the global regulatory requirements. Method development is an incessant and constant process where the objective is to improve the quality of the product [14].

ANALYTICAL METHOD VALIDATION

The general definition of validation is an establishment of documented evidence which provides a high degree and extent of assurance that a specific procedure, process, equipment, activity or system will consistently produce a product which meets its predetermined specifications and quality attributes.

Validation is closely related to the quality of the results; therefore, it is a significant and vital feature after the design and development of any analytical method. It is mandatory for each analytical method, either qualitative or quantitative, to be validated. The degrees of validation vary for the type of method employed and its subsequent application.

The method validation studies, guidelines and procedures have focused persistently and mainly on quantitative methods of analysis. Thus, validation is an imperative, significant and vital activity in the profiling of impurities, where the developed analytical method is used for the determination of GIs in drug compounds/substances is validated to ascertain easily that the method is apposite, appropriate and suitable for its intended and expected purpose. The analytical methods are validated with the following parameters.

LINEARITY AND RANGE

The linearity of an analytical method is its ability to draw out and deduce the test results that are directly proportional and comparative to the concentration of the analyte in the test samples within a given range. The linearity of an analytical method developed is verified by mathematical and statistical treatment of test results which are obtained by the analysis of samples with the analyte concentration across the averred range. Range of an analytical method developed is defined as the hiatus and interval between the upper and lower levels of analyte.

ACCURACY

The accuracy of a measurement system is defined as the degree of closeness of the results, that is, measurements of a quantity to that of actual (true) value. Therefore, the accuracy of an analytical method is the closeness of the test result attained by that method to that of the true value. Accuracy may be expressed in the form of percentage recovery by performing the assay procedure of the known amount of added analyte.

PRECISION

The precision of a measurement system is also referred to as reproducibility or repeatability. Thus, it is defined as the extent to which reiterated measurements under unaffected conditions show the unamended results. Precision is, thus, the degree of conformity and harmony among individual test results when the procedure is functional and repeatedly applied to multiple samples of a consistent sample. Precision is usually expressed in the form of the standard deviation or relative standard deviation.

SENSITIVITY AND SPECIFICITY

Sensitivity is the capability of the test procedure to determine diminutive disparity in concentration. For spectrophotometry, sensitivity is calculated in terms of Sandell's sensitivity (л). The sensitivity of the method heavily depends upon the investigational conditions. The information and acquaintance of sensitivity of the effect are imperative and essential, and an effortless detectable alteration in the intensity must be acquired by small changes in concentration.

Sensitivity is, in general, a statistical measure of the recital of a dual classification test, also known in statistics terminology as the classification function. Sensitivity is also called recall rate in some fields is thus the measure of the fraction of actual positives, which are appropriately recognised as such. An example is the percentage of the ailing populace who are acceptably recognised as having the condition.

Sandell's sensitivity is calculated as below:

$$\pi = \text{Conc.}(\mu g/100 \text{ ml}) \times 0.001/D^1 \text{ value(abs)}$$

The unit of Sandell's sensitivity: $\mu g/cm^2/0.001$ AU

The specificity of an analytical method is the ability of a method to measure precisely and explicitly the analytes in the existence of constituents that may be anticipated and predictable to be present in the mock-up mould.

ROBUSTNESS AND RUGGEDNESS

The ruggedness of an analytical method developed is the degree of the reproducibility of test results which are obtained by the investigation under diverse normal test conditions, such as dissimilar laboratories, diverse analysts, dissimilar instruments, diverse reagent, different elapsed assay times, diverse temperatures, unlike days, etc.

The robustness is, thus, the measure of the capability of an assay to stay unconcerned by small but premeditated and intentional variations in method parameters, which will provide an indication and suggestion of its reliability in regular usage. Dilapidation and differences in chromatography columns, mobile phases and insufficient method development are common causes of lack of robustness.

These parameters are consistent with the ICH Harmonized Tripartite Guidelines [15,16].

CONCLUSION

The present chapter describes and ascertains the minute and illustrated details regarding GIs in drug substances/compounds and drug products. The chapter also presented a terse discussion and provided invaluable information regarding the classification of impurities and the subsequent regulatory guidelines to control these impurities. Various analytical techniques have been developed and used in the isolation of these impurities. Thus, it was revealed that analytical techniques are required for the determination, fortitude and quantification of GIs along with method development and its validation.

REFERENCES

1. ICH S2 (R1). 2011. The tripartite harmonised ICH Guideline S2 (R1): Guidance on genotoxicity testing and data interpretation for pharmaceuticals intended for human use. 1–2.
2. Custer, L., Sweder, K. 2008. The role of genetic toxicology in drug discovery and optimization. *Curr Drug Metabol* 9:978–985.
3. Yadav, U., Dhiman, P., Malik, N., et al. 2013. Genotoxic impurities an overview. *J Biomed Pharm Res* 2:39–47.
4. Szekely, G. Amores de Sousa, M.C., Gil, M. 2015. Genotoxic impurities in pharmaceutical manufacturing: Sources, regulations, and mitigation. *Chem Rev* 115:8182–8229.
5. Derek, R. 2010. Control of genotoxic impurities in active pharmaceutical ingredients: A review and perspective. *Org Process Res Dev* 14:946–959.
6. David, J., McGovern, T. 2007 Toxicological overview of impurities in pharmaceutical products. *Adv Drug Deliv Rev* 59:38–42.
7. Committee for Medicinal Products for Human Use. 2006. *Guideline on the Limits of Genotoxic Impurities*. European Medicines Agency (EMEA), London, UK.
8. Genotoxic and Carcinogenic Impurities in Drug Substances and Products. 2008. US Department of Health and Human Services, Rockville, MD.
9. Guideline, ICH Harmonised Tripartite. 1995. Guidance on specific aspects of regulatory genotoxicity tests for pharmaceuticals S2A.
10. International Conference on Harmonisation (ICH). 2017. Assessment and control of DNA reactive (mutagenic) impurities in pharmaceuticals to limit potential carcinogenic risk, M7.
11. Kroes, R., Kleiner, J., Renwick, A. 2005. The threshold of toxicological concern concept in risk assessment. *Toxicol Sci* 86:226–230.
12. Ahuja, S., Alsante, K. 2003. *Handbook of Isolation and Characterization of Impurities in Pharmaceuticals*, Vol. 5. Academic Press, Cambridge, MA.
13. Ahuja, S., Scypinski, S. 2010. *Handbook of Modern Pharmaceutical Analysis*, Vol. 10. Academic Press, Cambridge, MA.
14. Yuri, V., Lobrutto, R. 2006. *HPLC for Pharmaceutical Scientists*. John Wiley & Sons, Hoboken, NJ.
15. Guideline, ICH Harmonised Tripartite. 2005. Validation of analytical procedures: Text and methodology Q2 (R1). *International Conference on Harmonization*, Geneva, Switzerland. 1–13.
16. Chow, C., Lee, Y., Lam, H., Zhang, X. 2004. *Analytical Method Validation and Instrument Performance Verification*. John Wiley & Sons, Hoboken, NJ.

14 Genotoxicity and Carcinogenicity Predictive Software Systems

INTRODUCTION

Software tools for the prediction of genotoxicity and carcinogenicity
Predictions have been done using a series of software tools, including

- A tool based on connoisseur rules (Derek v.12)
- Tools based on statistical methodology (CAESAR, Lazar, TOPKAT v. 6.2, HazardExpert [Pallas v 3.3.2.4]
- The previously named ToxBoxes [now called ACDToxSuite])
- A hybrid tool (Toxtree v.1.60).

In every case, a format for inferring the model outcomes in stipulations of categorical activities is either adopted or devised. The predictive feats of individual software tools have been appraised and contrasted by employing internal and external datasets.

The selection criteria for the above-mentioned software tools were on:

- Practical grounds
- Taking into account the in-house availability of software
- Budgetary checks
- Procurement constraints for the possession of the latest licences

The OECD QSAR Toolbox has been utilised extensively and increasingly owing to its freely available resource. However, it is mostly utilised for clustering chemicals and to assist read-across, moderately than a tool used for pre-defined QSAR algorithms. The exclusion of a given software tool from the study does not imply that it is not capable. Nevertheless, within the limitations, a varied range of methodologies, both statistical and expert-based, have been applied.

All the software mentioned above have been discussed in brief for their applicability and predictive capabilities.

Toxtree

- Toxtree is a lithe, user-friendly and open-source application that situates chemicals into different categories and envisages a range of toxic effect employing decision tree approaches [1].
- Toxtree can be downloaded from:
  - JRC (http://ecb.jrc.ec.europa.eu/qsar/qsar-tools/index.php?c=zTOX-TREE) and
  - Sourceforge (https://sourceforge.net/projects/toxtree/)
- Toxtree has been developed by the JRC in alliance with diverse consultants, particularly Ideaconsult Ltd. (Sofia, Bulgaria).
- A vital feature of Toxtree is the apparent reporting of the reasoning fundamental for each prediction. Toxtree v 1.60 includes the classification schemes for:
  - Systemic toxicity (Cramer scheme and extended Cramer scheme)
  - Mutagenicity and carcinogenicity (Benigni Bossa rule base and the ToxMic rule base on the *in-vivo* micronucleus (MN) assay).
- The Cramer scheme is perhaps the most extensively used approach for structuring chemicals to infer and assess the threshold of toxicological concern (TTC) [2].
- The current version of Toxtree applies the TTC scheme, alerts for skin sensitisation alerts and SMARTCyp, which is a two-dimensional method for the prediction of cytochrome P450-mediated metabolism. SMARTCyp predicts sites in a molecule that are labile for metabolism by cytochrome P450.
- The carcinogenicity/mutagenicity forecast generated by Toxtree is based on a decision tree which implements the Benigni Bossa rules and rules for the *in-vivo* MN assay.
- In addition, Toxtree concerns and pertains for the following QSAR models to query the chemicals to fit into the classes of aromatic amines or alpha, beta-unsaturated aldehydes:
  - *QSAR6* – testing of mutagenic activity of aromatic amines in *Salmonella typhimurium* TA100 strain (Ames test).
  - *QSAR8* – testing of carcinogenic activity of the aromatic amines in rodents (summing up activity from rats and mice).
  - *QSAR 13* – testing of mutagenic activity of alpha, beta-unsaturated aldehydes in *Salmonella typhimurium* TA100 strain (Ames test).
- There are some exclusions in the application of these QSARs, namely, QSAR6 and QSAR8 in Toxtree, which are applicable for aromatic amines with the exception of aromatic amines bearing a sulphonic group on the same ring; and QSAR13 applies to alpha, beta-unsaturated aldehydes excluding cyclic alpha, beta-unsaturated aldehydes [2].
- For the ultimate obligation of genotoxicity and carcinogenicity predictions, the weight of evidence has been applied.
- Overall, the QSAR analyses provide an additional sophisticated and advanced assessment than the structural alerts. The productivity of the QSARs for carcinogenicity was given more attention than the presence of structural alerts for (non)genotoxic carcinogenicity.

- Thus, when these QSARs gave a negative result, despite the presence of structural alerts, the final prediction was treated as negative, which implicates that there is a lack of toxicity.
- However, in the case of genotoxicity, the QSAR output and the structural alerts for genotoxic carcinogenicity were allocated equivalent weight that if either an alert or the QSAR delivers a positive prediction, and the overall prediction is positive. Therefore, the outcome of a QSAR for Ames mutagenicity was regarded as a deficient prediction of mammalian genotoxicity [2].
- When applying Toxtree for the prediction of Ames mutagenicity, only the alerts for genotoxic carcinogenicity are utilised. However, when the concern is for the prediction of classified mutagens, both the genotoxic carcinogenicity alerts (Benigni Bossa rule base) and the *in-vivo* MN alerts (ToxMic rule base) have been taken into consideration.
- The structural rules in Toxtree are based on expert knowledge rather than statistical derivations from the training sets. However, the Benigni Bossa rule base includes some QSARs in toting up to the structure-based rules [2].

CAESAR

- CAESAR comprises a series of statistical models which have been developed within the EU-funded CAESAR project http://www.caesar-project.eu/ [3].
- The models have been highly implemented into open-source software and are made available for online use.
- Predictions can be made for five endpoints:
  - Mutagenicity (Ames)
  - Carcinogenicity
  - Developmental toxicity
  - Skin sensitisation
  - Bioconcentration factor
- The CAESAR prediction of mutagenicity is based on support vector machine (SVM) approach and the Kazius Bursi database (http://www.cheminformatics.org/datasets/bursi) [3].
- The SVM modelling is pursued by an "expert facility" filter based on Benigni Bossa rules; it applies to compounds which have been presumed as safe by SVM [3]. Thus, the filter combines two sets of structural alerts with dissimilar distinguishing features:
  - The former, that is the "sharp" one, has the capability to augment the prediction accuracy, thereby attempting a precise detection of misclassified false negatives (FN).
  - The latter, that is the "suspicious" one, maintains with the FN removal in a manner such that it does not conspicuously reduce the prediction accuracy by generating excessively many false positives as well.
  - Compounds selected out by the first checkpoint are thereby classified as "mutagenic", which means they are active.

- Those pulled out by the second are hereby classified as "suspicious", which means that they are equivocal.
- Unaffected ones are at last classified as "non-mutagenic", attributing them as inactive.
- The prophecy of carcinogenicity by CAESAR is executed based on a counter propagation artificial neural network (CP ANN), which has a classification model and compounds from the CPDB (http://potency.berkeley.edu/cpdb.html). The software output categorises the molecules as either "positive", which classifies them as active, or "non-positive", which addresses them as inactive, with the possibility of the compounds to articulate and convey the activity or inactivity equal to 0.5 [3].

Lazar

- Lazar is an open-source software program that puts together calculations and forecasts of toxicological endpoints, referred to as mutagenicity, human liver toxicity, rodent and hamster carcinogenicity and maximum recommended daily dose by analysing the structural fragments in a training set [4].
- It utilises the statistical algorithms for classification such as the k-nearest neighbours (k-NNs) and kernel models, and for regression it employs the multilinear regression and kernel models [4].
- In comparison to the traditional k-NN, Lazar treats the chemical similarities not in values but as toxicity reliant values, which has an advantage in capturing only those fragments which are applicable and pertinent for the toxic endpoint which is under investigation [4].
- Lazar performs automatic applicability domain estimation and thus provides a self-assurance index for each prediction, and is utilised without expert knowledge. Lazar runs under the Linux operating system and a web-based prototype is also generously accessible at http://lazar.in- silico.de/.
- The mutagenicity prophecies by Lazar are based on the following parameters:
    - A k-NN algorithm and two datasets: Kazius Bursi (http://www.chemin-formatics.org/datasets/bursi/) and the so-called benchmark data set for in silico prediction of Ames mutagenicity (http://ml.cs.tu-berlin.de/toxbenchmark/).
    - Each forecast is related and connected with prediction confidence which is depicted between 0 and 1, which provides information about the presence and absence of studied compounds within the applicability domain (AD) of the model.
    - The developer anticipated a confidence value of higher than 0.025 as a reasonable hard cut-off for compounds within the mentioned AD. "The accuracy of prediction shrinks with the confidence value".
- The Lazar algorithm works by constitution of an illustration based local model that eliminates the chemical being forecasted from its local training set. Thus, the envisaged compound is never in the model training set [4].

HAZARDEXPERT

- HazardExpert is a component of the Pallas software which is developed by CompuDrug; http://compudrug.com/.
- It easily and efficiently predicts the toxicity of organic compounds based on toxic fragments and calculation of the bioavailability parameters such as logP and pKa [5].
- It is a rule-based system with an open comprehension base, which is user-friendly and thereby allows the user to either expand or modify the data on which the toxicity assessment depends [5].
- It covers the subsequent endpoints which are relevant to the dietary toxicity consideration:
  - Carcinogenicity
  - Mutagenicity
  - Teratogenicity
  - Membrane irritation
  - Immunotoxicity
  - Neurotoxicity
- The inferences yielded by oncogenicity and mutagenicity predictions by HazardExpert are expressed as a relative percentage of the toxicity values [5].
- Based on the range of the inferences, the proposed classification of chemicals to express the oncogenic/mutagenic activity is as follows:
  - Highly probable
  - Probable
  - Uncertain
  - Not probable
- To distinguish the HazardExpert predictions with the inferences of other software tools, the interpretation and elucidation are done in the following sequence where highly probable and probable chemicals are designated as active, uncertain chemicals are designated as equivocal and not probable as not active [5].

DEREK

- Derek for Windows (DfW) software is a SAR-based system developed by Lhasa Ltd., which is a non-profit company and an educational charity; it is available at https://www.lhasalimited.org/.
- DfW restrains for over 50 alerts that cover a wide range of toxicological endpoints in humans as well as in other mammals and bacteria.
- An alert is a toxicophore defined as a substructure which is known to be accountable and held responsible for the toxicity. It is, thus, allied with literature, observations and exemplars [6].
- A salient feature of DfW is the apparent reporting of the reasoning underlying each prediction.
- All the rules in DfW are based on either a hypothesis which relates to the mechanism of action of the chemical class or on the observed empirical relationships [6].

- The information used in the expansion and improvement of rules is inclusive of the published information, facts, figures, numbers and proposition from toxicological experts in industries, regulatory bodies and academia [6].
- The toxicity prophecies are the result of two processes.
  - The program first confirms whether any alerts in the information base are matching toxicophores in the query structure.
  - The reasoning engine then assesses the likelihood of a structure being toxic [6].
- There are nine levels of confidence in this regard which are as follows:
  - Certain
  - Probable
  - Plausible
  - Equivocal
  - Doubted
  - Improbable
  - Impossible
- DfW can be incorporated with Lhasa's Meteor software, which makes prediction of fate, thereby providing predictions of toxicity for both parent compounds and their metabolites.
- Genotoxicity alerts in Derek is inclusive of alerts for mutagenicity same as in bacteria and mammals and alerts for chromosomal damage based on the *in-vitro* chromosomal aberration (CA) assay and including effects that do not engage unswerving DNA damage, including the inhibition of DNA synthesis/repair, spindle function disruption, reactive oxygen species generation (ROS), energy depletion, thiol reactivity, and intercalation [6].

ToxBoxes

- ToxBoxes, which is now referred to as ACD/Tox Suite and marketed by ACD/Labs and Pharma Algorithms, predicts diverse toxicity endpoints, which are inclusive of genotoxicity, CYP3A4 inhibition, hERG inhibition, rodent LD50, ER binding affinity, organ-specific health effects and irritation and aquatic toxicity.
- It can be downloaded from http://www.acdlabs.com/products/admet/tox/ [7].
- The forecasts are related to confidence hiatus and prospects, thereby endowing with a numerical expression for the prediction dependability [7].
- The software has as a feature of the capability to categorise, discover and visualise explicit and precise structural toxicophores, which gives an insight into which part of the molecule is responsible for the toxic effect. It also categorises the analogues from its training set, which increases the confidence in the prediction. The algorithms and datasets are not disclosed [7].
- The predictions of genotoxicity by ToxBoxes rely on the probability that the query compounds are to be genotoxic in Ames test. The training data used in the software originates from Chemical Carcinogenesis Research Information (CCRIS) and Genetic Toxicology Data Bank (GENE-TOX).

- In establishing the training set, a compound was considered to be as genotoxic if at least one of the Ames results was positive; and if the result is not so, then the compound was considered non-genotoxic. In case of inconsistent results from different assays, the data is evaluated by connoisseurs and in some cases labelled as indecisive [7].
- The final training set is standardised Ames genotoxicity values. The neural network model is built using structural wreckage as descriptors. Molecules were degrading into atomic and chain-based fragments (chains of interconnected atoms).
- This model predicts if the chemical structure is more than 75% covered by fragments in the training set, then, for each compound, the "possibility of positive Ames test" and the "Ames test reliability index" is calculated [7].
- The method advocated switching the prospect values into binomial ones, such as either actives or inactives. This has the following framework according to the following rules:
  - If the "Probability of positive Ames test" is superior to 0.7, then the compound is predicted to be a mutagen, which means it is active.
  - If the "Probability of positive Ames test" is less than 0.3, then the compound is predicted to be a non-mutagen, which means it is inactive.
  - If the "Probability of positive Ames test" is between 0.7 and 0.3, then the result is predicted to be as **equivocal**.
- The ToxBoxes training set is easy to get via its database. But it is not convenient to validate and substantiate the overlap with the test sets. This is because the database can only be checked chemical by chemical; it cannot be extorted or investigated in an automated manner [7].
- When the analysis is performed with ToxBoxes, the noticeable point is to avoid some errors in toxicity prediction. Thus, it is essential again to encode the input file [7].

REFERENCES

1. http://ecb.jrc.ec.europa.eu/qsar/qsar-tools/index.php?c=TOXTREE (accessed March 20, 2020).
2. https://sourceforge.net/projects/toxtree/ (accessed March 20, 2020).
3. http://www.caesar-project.eu/ (accessed March 20, 2020).
4. http://lazar.in-silico.de/ (accessed March 20, 2020).
5. http://compudrug.com/ (accessed March 20, 2020).
6. https://www.lhasalimited.org/ (accessed March 20, 2020).
7. http://www.acdlabs.com/products/admet/tox/ (accessed March 20, 2020).

15 Poultry Manure Induced DNA Damage

INTRODUCTION

Redolent compounds are emitted from buildings, animal feed surfaces, manure storage and other treatment units, fodder piles and a variety of other emission sources [1]. Emission of redolent compounds from chicken sheds leads to an odour crisis in the contiguous area, resulting in complaints from residents or has some ill health effects [2]. An obnoxious odour is associated with poultry manure, which is a result of a combination of approximately 150 different compounds, including aldehydes, amines, ammonia, hydrogen sulphide, mercaptans, sulphur compounds and esters [2–4]. Odour is mainly due to the products formed upon the decomposition of chicken faeces, feathers and other litter by certain aerobic and anaerobic microorganisms. Each of these sources mentioned above has a distinct emission profile, which is unpredictable and remains irregular during the day and throughout the year [1].

Therefore, an effectual, helpful and efficient odour removal technique needs to be developed to tackle the great challenge of their disposal. Previous studies have demonstrated viable treatment of poultry manure with the mineral microbial preparation, which has successfully reduced odorants/redolents from the feedstock. These odorants are ammonia, dimethylamine (DMA), trimethylamine (TMA), isobutyric acid and other redolent compounds. A novel microbial mineral preparation composed of perlite and bentonite with spray-dried microorganisms is being used for the active removal of odours from poultry manure [5,6].

Ammonia, DMA, TMA, indole, phenol and butyric acid are the most general and regular compounds present in poultry manure [7–9]. Their health effects on humans and animals are well-known, and several studies have been done in-depth [10].

To summarise, these redolent compounds with extended exposure can have the following health effects:

- Pain in mucosal membranes in the respiratory tract
- Tracheal irritation
- Air sac inflammation
- Conjunctivitis
- Dyspnoea
- Respiratory tract damage
- Reddening
- Corneal clouding
- Reduction in respiratory rate
- Central nervous system disturbances [10]

There is still inadequate and paucity of *in-vitro* data linked with the cytotoxicity and genotoxicity of poultry redolent compounds pertaining to the application of cell lines. Therefore, this chapter deals with the cytotoxicity assay procedures and the associated requirements of the above-mentioned redolent compounds by employing the MTT (3-(4,5-dimethylthiazol-2-yl)-2,5-diphenyltetrazolium bromide) and PrestoBlue assays using chicken Leghorn Male Hepatoma (LMH) cell line [10]. These protocols will serve as appropriate for the concentrations of the redolent compounds regarding genotoxicity testing. Investigation for the mechanisms of toxicity of above-mentioned compounds redolent compounds can also be examined if cytotoxicity is an adjunct to genotoxicity, which can pilot to transformations and cancer.

The aim of the present chapter is to devise protocols for the measurement of genotoxicity (the measurement of DNA damage by employing the Comet assay) and cytotoxicity (necrosis in lactate dehydrogenase assay, along with IC_{50} values) of the above-mentioned redolent compounds from poultry manure in the chicken LMH cell line *in vitro*. In addition, the microscopic examination of any morphological changes in the cellular nuclei can also be studied by staining (apoptosis or necrosis after fluorescent staining).

PROTOCOLS

MATERIALS AND METHODS

Chemicals: Ammonia, DMA, TMA, indole, phenol, butyric acid, Waymouth's medium with no supplements and sterilised filters (0.22 μM pore size filter, membrane solutions).

Note: All compounds need to be prepared freshly on the day of the experiment.

LMH CELL CULTURE

- The chicken liver hepatocellular carcinoma cell line, LMH, is to be used in the experiments as a model cell line.
- Target of redolent compounds is first on the respiratory tract, and from there they are transported with blood to other organs, and then to the liver, which then participates in their detoxification.
- The culturing of LMH cells were in the form of a monolayer as described earlier [10]. Shortly after that, it was again cultured in T75 flasks which were collagen-coated in a Waymouth's medium with 7.5% sodium bicarbonate, 10% heat-inactivated foetal bovine serum (FBS), 25 mM HEPES and antibiotics (100 IU/mL penicillin and 100 μg/mL streptomycin).
- The cells were then incubated in a CO_2 incubator at 37°C in 5% CO_2 for 7 days to reach 80% confluence. These confluent cells were then detached with TrypLE™ Express for 10 min at 37°C, while remaining suspended in sterile PBS, aspirated off the plastic flask, which was then centrifuged (182 × g, 5 min), decanted and then re-suspended in fresh medium.
- These cells were then ready to use if a minimum of 90% viability was obtained when tested by Trypan blue exclusion.

COMET ASSAY (SCGE—SINGLE CELL GEL ELECTROPHORESIS ASSAY) – PROTOCOL

- The finishing concentration of LMH cells in every sample to be tested was adjusted to 105 cells/mL.
- The cells in non-supplemented Waymouth's medium were then incubated with precise concentrations of each redolent compound at 37°C for 60 min.
- The absolute series of concentrations of the redolent compounds in culture were selected based on the available studies where IC50 values of these redolent compounds have been determined. According to literature, it is 0.001%–0.006% for DMA and ammonia, 0.001%–0.12% for TMA, 0.003%–0.5% for butyric acid and Indole and 0.0004%–0.1% for phenol [10].
- The IC_{50} values of the above-mentioned redolent compounds by MTT and PrestoBlue assays, which were 0.02%–0.08% for ammonia; 0.03%–0.06% for DMA; 0.02%–0.08% for TMA; 0.11%–0.32% for butyric acid; and 0.06% for indole, depending on the incubation time and the assay [10].
- The Comet assay was then performed under alkaline conditions (pH > 13), as per the procedure of Błasiak and Kowalik [11,12].
- In short, after their incubation, the cells were centrifuged ($182 \times g$, for 15 min, at 4°C). They were then decanted and suspended in 0.75% low melting point Agarose, which was layered on to slides pre-coated with 0.5% normal melting point Agarose and lysed at 4°C for 60 min in a buffer containing 2.5 M NaCl, 1% Triton X-100, 100 mM EDTA and 10 mM Tris, pH 10.
- These slides were then placed in an electrophoresis unit, and the DNA was allowed to unwind for 20 min within an electrophoretic solution containing 300 mM NaOH and 1 mM EDTA.
- Electrophoresis was then conducted at 4°C for 20 min at an electric field strength of 0.73 V/cm (300 mA).
- Finally, the slides were neutralised in distilled water, and staining was performed with 2.5 µg/mL propidium iodide, and covered with coverslips. The slides were then observed at 200× magnification under a fluorescence microscope connected to a video camera and a personal computer-based image analysis system (Lucia-Comet version 7.0).
- With the help of a video camera, 50–100 images were arbitrarily selected from each sample and the percentage of DNA in the Comet tail was measured accordingly.

LACTATE DEHYDROGENASE ACTIVITY (LDH) ASSAY – PROTOCOL

- The basis of LDH cytotoxicity assay is the leakage of LDH, a cytoplasmic enzyme from cells when the plasma membrane is injured. This assay is useful to detect necrosis as well [13].
- According to the experimental protocol, 1×10^4 LMH count cells are put in a complete culture medium, in each well of a 96-well plate coated with collagen.

- The cells were then incubated for 24 h at 37°C in 5% CO_2 to set aside for attachment to the collagen-coated surface.
- The following day, the medium was aspirated and 200 μL of apiece concentration of the test compound in Waymouth's medium without FBS, was added to each well in eight repeats.
- The control samples consist of cells without any test agent. The cells were then incubated in a CO_2 incubator at 37°C in 5% CO_2 for 24 h for DMA and 48 h for ammonia and TMA.
- The final test concentrations were selected based on IC_{50} values; 0.004–1.0% for DMA and ammonia and 0.004%–1.0% for TMA.
- The concentrations were investigated, and the times of incubation were chosen to detect IC_{50} values [10]. The viability of the cells was then measured by MTT and PrestoBlue assays.
- In this protocol, the cytotoxic redolent compounds can be examined, along with the check for the mechanism of cell damage (loss in cell membrane integrity and probably necrosis).
- The assay was then conducted with a Cytotoxicity Detection Kit[PLUS], as per the manufacturer's instructions.
- Three control samples were included: background control (assay medium), low control (untreated cells) and high control (maximum LDH release).
- For determining the experimental absorbance values, the average absorbance values of the eight repeat samples and controls were calculated and then subtracted from the absorbance values of the background control.
- The cytotoxicity was determined by the following formula:

cytotoxicity (%) = {(exp. Value-low control)/(high control-low control)} × 100

- The absorbance was detected at 490 nm using a microplate reader.
- Results were then presented as mean ± standard deviation (SD).
- The mean error of the method was up to 10%.

CALCULATION OF IC_{50} – PROTOCOL

- IC_{50} is the concentration of the test compound requisite to decrease the cell survival rate to 50% of the control. These values are used as an extent of cellular compassion to a given treatment.
- IC_{50} value is determined by the following formula:

$$IC_{50} = (A - B) / (A - A_1) \times (C_{A1} - C_A) + C_A,$$

where,
  A is a 50% decrease in viability;
  A is % of viability > B;
  A_1 is % viability < B;
  C_A is the concentration of the compound for A and
  C_{A1} is the concentration of the compound for A_1 [10,14].

FLUORESCENCE MICROSCOPIC ANALYSIS –PROTOCOL

- Nuclear changes were observed by employing eight-well Lab Tek™ Chamber Slides in LMH cells in the presence of test compounds.
- Proceeding for the culture, the slides were coated with collagen I, as per the instructions provided by the manufacturer.
- LMH cells were kernelled on to each well with a concentration of 2.5×10^5 cells per well. The cytotoxic redolent compounds were selected (Ammonia, DMA and TMA) to check for the mechanism of cell death, damage or apoptosis using DAPI (4′,6-diamidino-2-phenylindole) staining.
- The final test concentrations were determined, as per the literature it is 0.03% for ammonia, DMA and TMA, which are close to the IC_{50} values estimated by employing MTT assay after 48 h of incubation [10].
- After exposure, medium with compounds was removed, cells were washed with PBS (phosphate buffer saline; pH 7.2) and fixed with 80% ethanol (for 20 min at room temperature).
- After air-drying, the cells were stained with 1 μg/mL DAPI in the dark. The morphology of cells was then observed at 1000× magnification under a fluorescent microscope which is connected to a digital camera and analysed using NIS-elements BR 3.0 imaging software.

STATISTICAL ANALYSIS

- Comet assay data was analysed using two-way analysis of variance (ANOVA).
- A particular mode of interaction × time was used to evaluate the effects stimulated by the chemicals at this mode of interaction.
- Differences between samples with normal distribution were evaluated by Student's t-test.
- Both Student's t-test and ANOVA were performed using appropriate software. Significant differences were acknowledged at $p < 0.05$. The results were then expressed as mean ± standard error of the mean in case of Comet assays and ± SD for LDH assay.

Note: Non-exposed cells (negative control)-induced DNA damage will be lesser while cell treatment with positive control (20 μM H_2O_2) will result in heavy DNA breakage.

GENERAL NOTATIONS

- Ammonia, DMA, TMA and butyric acid increase tail DNA in a dose-dependent manner at all concentrations.
- Ammonia and DMA are highly genotoxic.
- Ammonia, DMA and TMA can induce mild and moderate genotoxicity in LMH cells.
- Butyric acid induces extensive DNA damage at all concentrations.

- In contrast, phenol and indole highly increase tail DNA independent of the concentration.
- Their genotoxicity has a fluctuating effect.
- Higher doses of both these compounds results in complete DNA fragmentation in many cells, which can be due to apoptosis, but not true DNA damage.
- In addition, the number of comets per slide is lower for lower concentrations and unexposed cells.
- Thus, there is strong cytotoxicity of these compounds on the cells after an exposure of 60 min.

DISCUSSION

Redolent compounds generated from chicken sheds can harm both the environment as well as the health of humans and animals. There is a paucity of information regarding *in-vitro* studies for investigating genotoxic and cytotoxic mechanisms of poultry-related redolent compounds. The Comet assay determines the mechanisms of action of the test redolent compounds. The assay is proficient in detecting even low levels of DNA damage in a cell [15]. It can detect a diverse range of DNA damage, such as single and double-strand splits, base damage, interstrand or DNA protein cross-links, alkali-prone sites, oxidised purines and pyrimidines, as well as general DNA fragmentation [16].

The test concentrations must range from the highest acceptable cytotoxicity to little or no cytotoxicity because DNA damage is related and linked with cell decease [15]. In the existing discussion, the concentrations based on the previously available research was the basis for the investigation of the cytotoxic activity of test redolent compounds by employing MTT assay [10].

In the alkaline Comet assay, all redolent compounds provoke heavy damage to DNA in LMH cells. Butyric acid, phenol and indole elicit the most widespread genotoxic effects to the cells at a similar level. This materialises their cytotoxicity to cells in suspension and reticence of the DNA repair system (repair system has been besieged by extensive damage). In contrast, ammonia, DMA or TMA induce considerably more damage to DNA at higher concentrations.

When treated with a noxious compound, living cells may either impede their growing and dividing property, or die through necrosis or apoptosis. In general, cells undergoing necrosis swell and loose membrane veracity, due to which they lockdown and liberate their intracellular contents extracellularly. In contrast, cells undergoing apoptosis are characterised by the attenuation of the cytoplasm, chromatin condensation and DNA fragmentation [13].

Lactate dehydrogenase (LD) is an enzyme found within every living cell. A tetrazolium salt is used in LDH leakage assay. When cell membrane veracity is at conciliation, the presence of LD enzyme in the culture medium can be utilised as an indication of cell death.

- In the initial stage, LD produces reduced NADH (nicotinamide adenine dinucleotide), which is then catalysed to pyruvate by the oxidation of lactate.

- In the subsequent stage, a tetrazolium salt is transformed and converted to a coloured formazan by this newly synthesised NADH by an electron acceptor. The quantity of formazan can be computed by standard spectroscopy.
- The assay enumerates and records the percentage of necrotic cells in the sample. Unfortunately, the LDH release assay does not distinguish between primary necrosis and secondary necrosis as a corollary of apoptosis [13].

The most cytotoxic compound as per the LDH assay is DMA. Ammonia also depicted high cytotoxicity. The least cytotoxic substance was TMA. Similar results were reported for ammonia, DMA and TMA [10] using the MTT (3-[4,5-dimethylth iazol-2-yl]-2,5-diphenyltetrazolium bromide) and PrestoBlue assays. They are more toxic in the MTT assay.

Thus, it can be generalised that *"These disparities highlight the discrepancies in the mechanisms of cytotoxicity of these redolent compounds and the different nature of each test"*.

The LDH leakage assay is based on the release of the enzyme into the culture medium after cell membrane damage (cell death), whereas the MTT assay is based on the enzymatic conversion of MTT in the mitochondria. The formazan product which is formed is impermeable to the cell membranes, and therefore, it accumulates in healthy cells [17]. Thus, we can conclude, that *"these redolent compounds are critical and stern for the powerhouse of the cell, that is, mitochondria"*.

DAPI staining is a quick and expedient assay for apoptosis which is based on fluorescent detection. DAPI is an explicit DNA binding dye and is not completely permeable. Normal cells confer slight blue fluorescence with a round nucleus when stained uniformly. Along with the apoptosis process, the possibility of permeability increases, and the apoptotic cells produce blue fluorescence. The nucleus margination is observed, and the condensation of chromatin is easily stained. Cell contraction, chromatin condensation, nuclear division and nucleus margination are allied with the apoptotic mode of cell death.

The cytotoxic effect of the test redolent compounds is having the support of microscopic observations of nuclear morphological changes. Cells incubated with ammonia depict nuclei with either apoptotic (apoptotic bodies) or necrotic morphology, with nucleus fragmentation of different sizes.

Genotoxic activity of ammonia using the comet assay revealed that it is concentration dependent, and ammonia induced the DNA dent in chicken cells. In contrast, it was revealed that [18] ammonia induces micronuclei formation in Swiss albino mice and subdued DNA repair in mouse fibroblasts. The genotoxic effect of ammonia in humans [18] by analysis of blood samples of individuals exposed to ammonia (working in a fertiliser factory) and control individuals not exposed to ammonia revealed that there is augmented frequency of chromosomal aberrations, sister chromatid exchanges and an amplified mitotic index in blood samples of the exposed individuals. Thus, *"the incidence of DNA dent is increased along with the increased length of exposure"*.

It was also found that [19] Ammonia acts as an anti-mitotic agent, independent of the pH, and that it changes the morphology of the nuclei in LMH chicken cells. *In-vivo* animal experimentation with ammonia [19] depicted that excessive amounts of redolent compounds alter the colonic epithelial cell morphology, which increases cell proliferation.

It was also revealed that [20] the toxicity of ammonia present in urea and chicken manure, on plant roots, inclusive of wheat, canola, and faba, which widened and enlarged the toxicity zones was initiated at the root apex. There is progressive necrosis and attenuation of the root axis and root hairs. Ammonia is also responsible for the inhibition of root elongation. The toxic effect of poultry litter aqueous leachate [21] is due to ammonia and other anionic organic compounds. The characteristic apoptosis symptoms are observed after exposure to ammonia and the DAPI staining of LMH cells. It was also revealed that [22–27] ammonia can stimulate cell death due to apoptosis, and that the noxious upshots caused by ammonia are diverse and depend on the cell line type.

In the Comet assay, DMA and TMA lead to increased DNA damage in a dose-dependent manner [28]. DMA induces the conversion of gene and point reverse mutation following metabolic activation [29–33].

Phenol and indole induce strong DNA damage, independent of the concentration. Phenol was reviewed as a genotoxin and depicted that it causes a momentous boost in micronucleus frequencies when compared to the solvent control. It also induces a considerable increase in DNA damage as in the Comet assay on three different bio-materials, namely, lymphocyte of human, spermatid of the mouse and akaryocyte of crucian, which is statistically significant.

Phenol caused mutagenesis *in vitro* in *Vicia faba* and mouse spleen cells (sister chromatid exchange) [34,35], as well as induced DNA oxidative damage in human promyelocytic HL-60 and mouse bone-marrow cells [36]. Phenol tempts sister chromatid exchange in human lymphocytes [37,38]. Thus it can be generalised that it causes genotoxicity in a micronucleus test and is termed as genotoxic [39,40].

Butyric acid induces heavy and extensive DNA damage. This ability of butyric acid to reduce cell growth, followed by the increase in apoptosis [41–42], is an indication that butyric acid has bimodal effects on cell propagation and its survival. Apoptosis is tempted by high levels of butyric acid which can induce apoptosis in different cell lines *in vitro* [43–45].

Low concentrations of ammonia and TMA exhibit either little or no cytotoxicity in the LDH assay, but they are highly genotoxic in comparison with the Comet assay. The similar effect is noted in the LDH assay for all tested redolent compounds as per the MTT assay [10]. Thus, it can be concluded that "*a cytotoxic agent may be genotoxic (cause damage to DNA), or may be disparaging for cell organelles (e.g. cell membrane or mitochondria), but every cytotoxic chemical does not induce genotoxicity*".

Furthermore, it can be concluded that "*a genotoxic substance can induce DNA dent at non-cytotoxic concentrations*". Genotoxicity can direct cancer-causing mutations, but it cannot be cytotoxic for the cell. Thus, Ammonia, DMA and TMA are very harmful to LMH cells, possibly because there exist various similarities in their chemical structures; therefore, cellular interactions lead to heavy downstream toxicity.

CONCLUSION

In conclusion, redolent compounds tend to be cytotoxic and genotoxic *in vitro*, and hence, they are also likely to be harmful *in vivo*. Ammonia, DMA, TMA, phenol, indole and butyric acid have extensive genotoxic and cytotoxic profile, which can easily cause changes in the pH of the medium. Ammonia, DMA and TMA induce cytotoxicity in LMH chicken cell line by loss of cell membrane veracity and, most likely, cell necrosis. Moreover, they alter the nuclear morphology of LMH cells, causing chromatin condensation and fragmentation, which can also boost apoptotic bodies, as observed with DAPI staining. Thus, these compounds can provoke cell fatality in two ways: necrosis and apoptosis. The information presented in this chapter further needs to be confirmed by performing additional *in-vitro* studies and tests for detection of apoptosis. Finally, the Comet Assay and LDH Assay, along with DAPI staining, are suitable, apposite, receptive and sensitive methods to screen and select redolent compounds for genotoxicity and cytotoxicity testing.

REFERENCES

1. Jacobson, L., Bicudo, J., Schmidt, D., et al. 2003. Air emissions from animal production buildings. Available online: http://www.isah-soc.org/userfiles/downloads/proceedings/2003/mainspeakers/18JacobsonUSA.pdf (accessed 10 March 2020).
2. Dunlop, M., Blackall, P., Stuetz, R. 2016. Odour emissions from poultry litter—A review litter properties, odour formation and odorant emissions from porous materials. *J Environ Manag* 177:306–319.
3. Li, H., Xin, H., Burns, R., et al. Air emissions from Tom and Hen Turkey Houses in the U.S. Midwest. Available online: http://www.abe.iastate.edu/adl/files/2011/10/Turkey-Emissions1.pdf (accessed 20 March 2020).
4. Rabadheera, C., Mcconchie, R. Phan-Thien, K. 2017. Strategies for eliminating chicken odour in horticultural applications. *World Poult Sci J* 73:365–378.
5. Gutarowska, B., Borowski, S., Durka, K. et al. 2009. Screening of microorganisms capable to remove odorous compounds from poultry manure. *Przem Chem* 88: 440–445.
6. Borowski, S., Matusiak, K., Powałowski, S. 2017. A novel microbial mineral preparation for the removal of offensive odors from poultry manure. *Int Biodeter Biodeg* 119:299–308.
7. McGahan, E. 2002. Strategies to reduce odour emissions from meat chicken farms. *Proc Poult Inf Exch.* 27–39. Available online: http://citeseerx.ist.psu.edu/viewdoc/download?doi=10.1.1.582.29&rep=rep1&type=pdf (accessed on 3 March 2020).
8. Dinh, H. Analysis of ammonia and volatile organic amine emissions in a confined poultry facility. Available online: http://digitalcommons.usu.edu/etd/598/ (accessed 3 March 2020).
9. Gutarowska, B., Matusiak, K., Borowski, S., et al. 2014. Removal of odorous compounds from poultry manure by microorganisms on perlite bentonite carrier. *J Environ Manag* 141:70–76.
10. Nowak, A., Matusiak, K., Borowski, S. 2016. Cytotoxicity of odorous compounds from poultry manure. *Int J Environ Res Public Health* 13:1046.
11. Błasiak, J., Kowalik, J. 2000. A comparison of the in vitro genotoxicity of tri- and hexavalent chromium. *Mutat Res* 469:135–145.

12. Nowak, A., Śliżewska, K. 2014. β-Glucuronidase and β-glucosidase activity and human fecal water genotoxicity in the presence of probiotic lactobacilli and the heterocyclic aromatic amine IQ in vitro. *Environ Toxicol Pharmacol* 37:66–73.

13. Chan, F., Moriwaki, K., De Rosa, M. 2013. Detection of necrosis by release of lactate dehydrogenase (LDH) activity methods. *Mol Biol* 979:65–70.

14. OECD Guidelines for the Testing of Chemicals, Section 4. Test No. 442D: In Vitro Skin Sensitisation are-nrf2 Luciferase Test Method. Available online: https://ntp.niehs.nih. gov/iccvam/suppdocs/feddocs/oecd/oecd-tg442d-508.pdf (accessed 20 March 2020).

15. Tice, R., Agurell, E., Anderson, D., et al. 2000. Single cell gel/comet assay: Guidelines for in vitro and in vivo genetic toxicology testing. *Environ Mol Mutagen* 35:206–221.

16. Collins, A. 2004. The comet assay for DNA damage and repair. *Mol Biotechnol* 26: 249–261.

17. Fotakis, G., Timbrell, J. 2006. In vitro cytotoxity assays: Comparison of LDH, neutral red, MTT and protein assay in hepatoma cell lines following exposure to cadmium chloride. *Toxicol Lett* 160:171–177.

18. Yadav, J., Kaushik, V. 1997. Genotoxic effect of ammonia exposure on workers in a fertility factory. *Indian J Exp Biol* 35:487–4920.

19. Mouillé, B., Delpal, S., Mayeur, C. 2003. Inhibition of human colon carcinoma cell growth by ammonia: A non cytotoxic process associated with polyamine synthesis reduction. *Biochim Biophys Acta* 1624:88–97.

20. Pan, W., Madsen, I., Bolton, R., et al. 2016. Ammonia / Ammonium toxicity root symptoms induced by inorganic and organic fertilizers and placement. *Agron* 108:2485–2492.

21. Gupta, G., Borowiec, J., Okoh, J. 1997. Toxicity identification of poultry litter aqueous leachate. *Poult Sci* 76:1364–1367.

22. Galitskaya, P., Selivanovskaya, S. 2016. Co composting as a method to decrease toxicity of chicken manure. *Int J Adv Biotechnol Res* 7:1276–1282.

23. Delgado, M., Martin, J., De Imperial, R. 2010. Phytotoxicity of uncomposted and composted poultry manure. *Afr J Plant Sci* 4:154–162.

24. Gupta, G., Kelly, P. 1992. Poultry litter toxicity comparison from various bioassays. *J Environ Sci Health Part A Environ Sci Eng Toxicol* 27:1083–1093.

25. Sli,vac, I., Blajić, V., Radošević, K., et al. 2010. Influence of different ammonium, lactate and glutamine concentrations on CCO cell growth. *Cytotechnology* 62:585–594.

26. Tae, K., Lee, G. 2015. Glutamine substitution: The role it can play to enhance therapeutic protein production. *Pharm Bioprocess* 3:249–261.

27. Mirabet, M., Navarro, A., Lopez, A. 1997. Ammonium toxicity in different cell lines. *Biotechnol Bioeng* 56:530–537.

28. Galli, A., Paolini, M., Lattanzi, G. 1993. Genotoxic and biochemical effects of dimethylamine. *Mutagenesis* 8:175–178.

29. Pool, B., Brendler, S., Liegibel, U., et al. 1990. Employment of adult mammalian primary in toxicology: In vivo and in vitro genotoxic effects of environmentally significant N nitrosodialkylamines in cells of the liver, lung, and kidney. *Environ Mol Mutagen* 15:24–35.

30. United States Environmental Protection Agency, Acute Exposure Guideline Levels (AEGLs) for Dimethylamine. Available online: https://www.epa.gov/sites/production/files/2014-08/documents/dimethylamine_tsd_interim_version_106_2008.pdf (accessed 4 March 2020).

31. Mortelmans, K., Haworth, S., Lawlor, T., et al. 1986. Salmonella mutagenicity tests. 2. Results from the testing of 270 chemicals. *Environ Mutagen* 8:1–119.

32. MHLW (Japanese Ministry of Health, Labour and Welfare). 2003. *Trimethylamine*. MHLW, Tokyo, Japan.

33. MHLW (Japanese Ministry of Health, Labour and Welfare). 2003. *In Vitro Chromosomal Aberration Test of Trimethylamine on Cultured Chinese Hamster Cells.* MHLW, Tokyo, Japan.
34. Li, Y., Qu, M., Sun, L. 2005. Z. Genotoxicity study of phenol and o cresol using the micronucleus test and the comet assay. *Toxicol Environ Chem* 87:365–372.
35. Zhang, Z., Yang, J., Zhang, Q., et al. 1991. Studies on the utilization of a plant SCE test in detecting potential mutagenic agents. *Mutat Res* 261:69–73.
36. Kolachana, P., Subrahmanyam, V., Meyer, K., et al. 1993. Benzene and its phenolic metabolites produce oxidative damage in HL60 cells in vitro and in the bone marrow in vivo. *Cancer Res* 53:1023–1026.
37. Morimoto, K., Wolff, S., Koizumi, A. 1983. Induction of sister chromatid exchanges in human lymphocytes by microsomal activation of benzene metabolites. *Mutat Res* 119:355–360.
38. Erexson, G., Wilmer, J., Kligerman, A. 1985. Sister chromatid exchanges induction in human lymphocytes exposed to benzene and its metabolites in vitro. *Cancer Res* 45:2471–2477.
39. Ciranni, R., Barale, R., Marrazzini, A., et al. 1988. Benzene and the genotoxicity of its metabolites: I Transplacental activity in mouse fetuses and in their dams. *Mutat Res* 208:61–67.
40. Chen, H., Eastmond, D. 1995. Synergistic increase in chromosomal breakage within the euchromatin induced by an interaction of the benzene metabolites phenol and hydroquinone on mice. *Carcinogenesis* 16:1963–1969.
41. Reddy, M., Storer, R., Laws, G., et al. 2002. Genotoxicity of naturally occurring indole compounds: Correlation between covalent DNA binding and other genotoxicity tests. *Environ Mol Mutagen* 40:1–17.
42. Kurita Ochiai, T., Hashizume, T., Yonezawa, H., et al. 2006. Characterization of the effects of butyric acid on cell proliferation, cell cycle distribution and apoptosis. *FEMS Immunol Med Microbiol* 47:67–74.
43. Watkins, S., Carter, L., Mak, J. 1999. Butyric acid and tributyrin induce apoptosis in human hepatic tumour cells. *J Dairy Res* 66:559–567.
44. Kurita Ochiai, T., Ochiai, K., Fukushima, K. 2001. Butyric Acid induced T cell apoptosis is mediated by caspase -8 and -9 activation in a fas - independent manner. *Clin Diagn Lab Immunol* 8:325–332.
45. Kurita Ochiai, T., Seto, S., Suzuki, N., et al. 2008. Butyric acid induces apoptosis in inflamed fibroblasts. *J Dent Res* 87:51–55.

16 Expert Opinion on the Basis of the Content and Easy Understandable Rundown for the Reader

INTRODUCTION

A genotoxic agent is a drug or a chemical which is responsible to cause aberrations or mutations in the DNA structure and which may lead to cancer. These agents act by altering the chromosomal structures, forming rings, breaks, joins, etc., and can be identified by the chromosomal aberration test. Any drug that prevents the genotoxic effect of a clastogenic agent is referred as an anti-clastogenic or anti-mutagenic agent. The major management of genotoxicity deals with:

- DNA repair methods
- Metabolism of harmful chemical clastogens
- Utilisation of anti-cancer drugs

The drugs used for treatment and management of genotoxicity also act as anti-cancer agents, for example, alkylating agents, intercalating agents and enzyme inhibitors. Certain plant extracts such as flavonoids and ellagic acid are found to seize pharmacological activity and are used as anti-mutagenic agents.

Competent removal of DNA damage from the transcribed sequences leads to the enhancement of cell survival as it enables the cells to express vital genes before complete removal of all damaged DNA. Repairing of damaged DNA is auxiliary targeted within the transcriptionally active strand of the gene, though some damage arises on the non-transcribed strand, albeit at a low level. In addition to strand specificity, the repair rate is determined by a multipart interplay among:

- Adduct structure
- Accessibility to repair enzymes
- Ability to seize transcription and the DNA conformation

A consequence of DNA damage provides a brief review on the outcomes of DNA damage, such as effects on the nervous system, cardiovascular system, lymphatic system and various other disorders such as ageing, cancer and genome instability. Moreover, the diverse innate mechanisms of DNA repair with respect to specific DNA damage is taken into consideration.

Genotoxins interact with the DNA, causing mutations and damage to the structure which leads to cancer. They act by changing the chromosomal structure either by addition, deletion, duplication or forming rings. These mutations lead to various diseases including cancer. Therefore, it is of utmost importance to consider genotoxicity studies to avoid the potential damage caused by it. Genotoxicity tests are performed to identify whether a drug or substance has the potential to cause mutation and genotoxicity. This helped in identification of the hazards during the early stages of drug development. This will be of assistance in comprehending the mechanism of the mutation and genotoxicity, thereby creating improved and enhanced techniques and approaches to avert and avoid the frequency of mutation and genotoxicity.

Currently, drug discovery and development is speedy, time profitable and dynamic due to the use of pioneering technology, such as genomics, high-throughput screening and proteomics. Pharmaceutical companies and other regulatory agencies have to assemble the challenges of the 21st century, and toxicologists and other scientists need to re-evaluate the obtainable protocols in the changing environment.

The accessible guidelines, in a developing country like India, need to be re-evaluated and modified in accordance with growing globalisation. By adopting the ICH guidelines, the process of new drug approval can be rationalised.

Both expert and QSAR methods are uniformly successful in the prediction of the mutagenic activity of compounds on which they are being modelled, but when it comes to true external data sets, they fall short. Directed development and testing of mechanistic SARs for the compounds incorrectly predicted leads to considerable improvement in the predictive systems. Development of more mechanistic SARs can extrapolate *in-silico* models into areas where mutagenic activity has not been explored. Thus, more confidence in predictions can be achieved by the model outside the domain of applicability.

Although computational strategies are applied to chromosomal damage, it is not adequate to simply relate the current paradigm of QSAR model development. Therefore, there is a need to identify toxicological knowledge gaps and carry out focused testing so that a better understanding of the mechanistic SARs can be achieved or indeed biological pathways behind the observed toxicity.

The currently accessible *in-silico* loom to genotoxicity prediction appends a value to the drug development process. However, similarly, it is apparent that the existing tools do not perform up to the desired degrees of both sensitivity and specificity. Although noticeable and discernible progress in concordance can be made via expert interpretation of the out-of-the-box calls, it is comprehensible that each computational model still requires wide-ranging modification prior to the eventual objective of reinstating biological assays can be accomplished. Thus, it is anticipated that one area where noteworthy revisions can be made is in the reporting of non-bacterial genotoxic responses. Applying specific genotoxicity data applicable and relatable to non-covalent DNA interactions should show the way to a discernible amplification in the predictive assessment and significance.

Thus, it can be concluded that the *Allium cepa* test is an exceptional bioindicator of chromosomal alterations, as well as general genotoxicity. Currently, due to key concern with environmental pollution, the *Allium cepa* test has occupied an imperative role in the prevention and prediction of the environmental impact caused by the exercise and removal of substances, including drugs and herbicides.

Although the test is merely a primary appraisal of genotoxicity, significant scientific discoveries and new adaptations of the test might reveal copious possibilities of its use, which will help in avoiding the use of animals for testing. As the method improves, more augmentation and investigation will lead us to get the most use for the benefit of the planet.

The most common assays for testing genotoxicity of NMs have been described and discussed. In general, the assays used had some interference and drawbacks which were identified. For the Comet assay, a risk for overestimation of the DNA damage has been suggested when high concentrations of reactive NPs are tested *in vitro* because of the additional damage caused during the assay performance. Most likely, the NPs that cause additional damage also cause "real" damage, and thus the risk for "false positives" seems rather small.

Numerous test schemes and processes have been utilised and employed to review and evaluate the genotoxicity of NMs, with almost a comparable echelon of outcomes and up-shots.

Thus, a small amount of wrapping up on NM genotoxicity can be prepared, regardless of a considerable, significant and extensive volume of work. Through this chapter an attempt has been made to seek out the assessment and evaluation of the literature, with an observation regarding the commendations and counsel on corroborated and authenticated methods and systems for genotoxicity assessment of NMs.

Numerous topics and concerns have been documented from this analysis, which is inclusive of an extensive distinction in the physical and chemical properties of NMs, unswerving NM categorisation and depiction in the test medium, numerous test systems failing to meet the OECD standards, obscurity of applying NMs to biological systems which is inclusive of its uptake, meddling and prying of NM with the test endpoint, impending and prospective disparity in the systemic allocation *in vivo*, and be deficient in a definitive mechanism of action.

Based on the current data, NM genotoxicity responses are less significant than pragmatic conventional DNA detrimental agents, which depends on genotoxicity which is stimulated via a derivative due to a certain extent than an up-shot of undeviating DNA interaction.

As an approach to the fore, the following recommendations can be proposed:

- The utility of cautiously defined NMs is inclusive of the categorisation in the test medium.
- An appraisal and evaluation of uptake and allocation within cells and *in-vivo* systems.
- Dose assortment is carefully chosen to keep away from object interrelated to system overwork.
- A custom-made test battery that is inclusive of genotoxicity testing in *in-vitro* mammalian mutagenicity and chromosomal damage assays, combined with assay variation as described within.
- Observation of investigations and cell systems as illustrated and expressed in the OECD TG.
- A superior and better effort on the considerate mechanisms.

Although not much literature is available for the genotoxicity of herbals, some plants whose genotoxicity has been studied and reported are discussed in the coming section. In 2012, Kelber et al. compiled a review on the genotoxicity of herbal plants listing the names of plants and highlighting how the results were obtained after the study. In 2014, an experimental study was conducted for the genotoxicity assessment of *Valeriana officinalis* L., radix root. Sponchiado et al. in 2016 compiled a systemic review of medicinal plants for quantitative genotoxicity assay. They discussed and conferred that a choice of methods are available for the genotoxicity assay which can be employed and utilised for the evaluation of the genotoxicity potential of medicinal plant extracts. These methods are highly recommended by regulatory agencies.

Based on the findings, it was concluded that to conduct a thorough study about possible genotoxic effects of any medicinal plant, it is important to include bacterial and mammalian tests, with at least one *in-vivo* assay. Further, these tests should be capable of detecting outcomes such as mutation induction, clastogenic and aneugenic effects and structural chromosomal abnormalities.

Mélanie Poivre in 2017 summarised the genotoxicity of phytoconstituents isolated from various plants. Some of the contribution in research for genotoxicity of herbals were listed.

The concept of toxicity study for HMPs is not new. From ancient times, it has been based on the experience of people about how much, when and where to use herbal products for the treatment of ailments. Because of the adulteration and changes in climatic and geographical conditions, it is the need of the hour to explore genotoxicity, carcinogenicity, mutagenicity, adverse effects and side effects of all traditional herbal plants. Most of the secondary metabolites prove to be toxic at high doses. OECD, WHO, EMA and ICH have collaboratively discussed the detailed procedure for ensuring the safety of herbals through *in-vitro* and *in-vivo* toxicity study. These procedures will also help in maintaining data and records for future use of herbals.

The experience of the allopathic industry suggests that regulatory guidelines and their implementations are necessary to support science and quality of research. Time has come to accept the same for herbal medicinal products. The estimation of genetic biomarkers would help estimate the potential toxicity of medicinal herbs to regulate medicinal plant consumption, which would be an important measure of public health protection.

Genotoxic events are a crucial step in the initiation of cancer. To assess the risk of cancer, genotoxicity assays, including Comet, micronucleus, chromosomal aberration, bacterial reverse and sister chromatid exchange assay can be performed. Compared with *in-vitro* genotoxicity assay, *in-vivo* genotoxicity assay has been used to verify *in-vitro* assay results and provide biological significance for certain organs or cell types.

In 2009, ICH issued an absolute model paper in which an M7 topic for control of GTIs was agreed upon. This document listed the problems with interpreting the multiple current regulations that had evolved over the last decade, as well as the main issues for which clarifications and guidance were required. Therefore, it was proposed to move forward with work towards a new guideline, entitled "Assessment and Control of DNA Reactive (Mutagenic) Impurities in Pharmaceuticals to Limit Potential Carcinogenic Risk". This approach was authorised in 2010 and an expert working group (EWG) was established, which consisted of two members nominated by the six sponsors of the ICH, and one member nominated by Health Canada, WHO and EFTA as observers. This EWG in 2010 drafted guideline for comments in late 2012.

The guidance addressed the topics as listed below:

- Acceptable levels of GTIs during drug development
- Acceptable levels of GTIs for marketing
- Acceptable levels of GTIs with a strong likelihood to exhibit a threshold effect
- The merit of the TTC approach
- How to treat structurally related GTIs
- The data required to support higher daily intakes than allowed by the TTC
- Genotoxic degradants and what control strategy to expect and apply

The last 10 years have seen GTIs emerge as a noteworthy and innovative regulatory topic for the pharmaceutical industry. Therefore, it is expected that the planned issuance of the guidelines will further clarify and support an efficient regulatory path for GTIs during drug development and commercialisation of new APIs.

It should be noted that "concordance" is a measure which is used numerously for evaluating the performance of a test system. At present, it is the level of agreement between the results in the *in-vitro* tests relative to the expected *in-vivo* test outcomes for genotoxicity or carcinogenicity. Concordance relies profoundly on having reasonably equal numbers of carcinogens/genotoxins and non-carcinogens/non-genotoxins. Many prior concerted or validation trials have include large numbers of carcinogens but a very few non-carcinogens. It should also be noted that while all of the chemicals mentioned in this chapter are commercially available, chemical laboratories should be aware of the quality and must obtain certificates of analysis for the test compounds regarding the details of purity and impurities. For good preparation of test chemical solutions, sufficient care should be taken in the suitable choice of the solvent/vehicle. Moreover, careful handling of chemicals is imperative, especially when handling IARC group 1 chemicals or coded chemicals.

After careful contemplation of the available literature, and in particular of data published since 2008, the updation pertaining to the lists of chemicals that can be used in the evaluation of modified or new mammalian cell genotoxicity assays has been done. These lists arrange the chemicals according to the expected positive results *in vitro* or negative results, and the latter includes chemicals currently suspected of giving "deceptive" or "extraneous" positive results in existing assays. It has been noted that the chemicals that have been deleted since the publication are not necessarily considered to be "erroneous". However, they are considered to be less vigorous candidates for the groups that have been presented, and any data obtained should be viewed with some caution. It is, thus, widely anticipated that these lists may prolong and maintain to provide quite functional and efficient reference chemicals for scientists seeking to modify existing assays or introduce new assays.

The terse and exhaustive description ascertains the minute and illustrated details regarding genotoxic impurities (GIs) in drug substances/compounds and drug products. The matter also laid a terse discussion and provided invaluable information regarding the classification of impurities and the subsequent regulatory guidelines to control these impurities. Various analytical techniques have been developed and used in the isolation of these impurities. Thus, it was revealed that analytical techniques are required for their determination, fortitude and quantification along with method development and its validation.

Prediction were made using a series of software tools, including:

- A tool based on connoisseur rules (Derek v.12)
- Tools based on statistical methodology (CAESAR, Lazar, TOPKAT v. 6.2, HazardExpert [Pallas v 3.3.2.4])
- Previously named ToxBoxes (now called ACDToxSuite)
- A hybrid tool (Toxtree v.1.60).

In every case, a format for inferring the model outcomes in stipulations of categorical activities is either adopted or devised. The predictive features of the individual software tools have been appraised and contrasted by employing internal and external datasets.

The selection criteria for the above-mentioned software tools were on:

- Practical grounds
- Taking into account the in-house availability of software
- Budgetary checks
- Procurement constraints for the possession of the latest licences

The information has been provided about redolent compounds that they tend to be cytotoxic and genotoxic *in vitro*, and are hence likely to be harmful *in vivo*. Ammonia, DMA, TMA, phenol, indole and butyric acid have an extensive genotoxic and cytotoxic profile, which can easily cause changes in the pH of the medium. Ammonia, DMA and TMA induce cytotoxicity in LMH chicken cell line by loss in cell membrane veracity and, most likely, necrosis. Moreover, they alter the nuclear morphology of LMH cells, thereby causing chromatin condensation and fragmentation, which can also boost apoptotic bodies, as observed with DAPI staining. Thus, these compounds can provoke cell fatality in two ways: necrosis and apoptosis. The information perceived in this chapter further needs to be confirmed by performing some additional *in-vitro* studies and tests for detection of apoptosis. Finally, the Comet assay and LDH assay, along with DAPI staining, are suitable, apposite, receptive and sensitive methods to screen and select redolent compounds for genotoxicity and cytotoxicity testing.

The explanation and elucidation of authentic and factual problems of genotoxicity are principally based on the understanding of DNA damaging methods at a molecular level, subcellular stage, cellular point, organ level, system and organism echelons.

Contemporary approaches to scrutinise the human health effects should be custom-made to augment and boost their performance for supplementary unswerving results, as well as novel techniques such as toxicogenomics, epigenomics and single-cell advances must be amalgamated into genetic well-being evolutions. The investigated novel biomarkers by the omic modus operandi will provide strong genotoxicity evaluation to decrease the cancer risk.

Index

151